Safety Assurance for Environmental Introductions of Genetically-Engineered Organisms

NATO ASI Series

Advanced Science Institutes Series

A series presenting the results of activities sponsored by the NATO Science Committee, which aims at the dissemination of advanced scientific and technological knowledge, with a view to strengthening links between scientific communities.

The Series is published by an international board of publishers in conjunction with the NATO Scientific Affairs Division

A Life Sciences **B Physics**	Plenum Publishing Corporation London and New York
C Mathematical and Physical Sciences **D Behavioural and Social Sciences** **E Applied Sciences**	Kluwer Academic Publishers Dordrecht, Boston, and London
F Computer and Systems Sciences **G Ecological Sciences** **H Cell Biology**	Springer-Verlag Berlin Heidelberg New York London Paris Tokyo

Series G: Ecological Sciences Vol. 18

Safety Assurance for Environmental Introductions of Genetically-Engineered Organisms

Edited by

Joseph Fiksel

Teknowledge Inc.
1850 Embarcadero Road
P.O. Box 10119
Palo Alto, CA 94303, USA

Vincent T. Covello

School of Public Health
Columbia University in the City of New York
60 Haven Avenue, B-1
New York, NY 10032, USA

Springer-Verlag
Berlin Heidelberg New York London Paris Tokyo
Published in cooperation with NATO Scientific Affairs Division

Proceedings of the NATO Advanced Research Workshop on Safety Assurance for Environmental Introductions of Genetically-Engineered Organisms held in Rome, Italy, June 6–10, 1987

Library of Congress Cataloging-in-Publication Data. NATO Advanced Research Workshop on Risk Analysis a Approaches for Environmental Introductions of Genetically-Engineered Organisms (1987 : Venice, Italy) Safety assurance for environmental introductions of genetically-engineered organisms / edited by Joseph Fiksel, Vincent T. Covello. p. cm.–(NATO ASI series. Series G, Ecological sciences ; vol. G 18) "Proceedings of the NATO Advanced Research Workshop on Risk Analysis Approaches for Environmental Introductions of Genetically-Engineered Organisms held in Venice, Italy, June 6–10, 1987"–CIP t.p. verso. "Published in cooperation with NATO Scientific Affairs Division."

ISBN-13: 978-3-642-73171-6 e-ISBN-13: 978-3-642-73169-3

DOI: 10.1007/978-3-642-73169-3

1. Biotechnology–Safety measures–Congresses. 2. Genetic engineering–Safety measures–Congresses. I. Fiksel, Joseph R. II. Covello, Vincent T. III. North Atlantic Treaty Organization. Scientific Affairs Division. IV. Title. V. Series: NATO ASI series. Series G, Ecological sciences ; vol. 18.

TP248.14.N37 1987 363.17–dc 19 88-39157

Softcover reprint of the hardcover 1st edition 1988

2131/3140-543210 – Printed on acid-free paper

ACKNOWLEDGEMENTS

This publication and the NATO Workshop upon which it was based were made possible through the financial and administrative support of the following organizations:

North Atlantic Treaty Organization (NATO)
National Science Foundation, U.S.A.
International Life Sciences Institute / Risk Science Institute
Society for Risk Analysis

The authors wish to express their appreciation for the energy and enthusiasm shown by all Workshop participants, and their special thanks to Diane Guysenir, Carol Mandel, and Christina Huemer for their help in organizing the Workshop.

TABLE OF CONTENTS

P A R T I

WORKSHOP SUMMARY

Workshop Summary:
Recommendations for a Scientific Approach to Safety Assurance for Environmental Introductions of Genetically-Engineered Organisms

Preface

This report summarizes the consensus of an international group of 35 scientific and regulatory experts who participated in a NATO workshop in June 1987, co-sponsored by the National Science Foundation, ILSI-Risk Science Institute, and the Society for Risk Analysis. The objectives of the workshop were:

- to evaluate the available safety assurance and risk assessment methods for environmental introductions of genetically-engineered organisms,
- to establish a basis for improving the available methods, and
- to develop principles for conducting scientific assessments.

The conclusions expressed here represent the professional views of the individual participants, and do not reflect the policies or positions of the institutions with which they are affiliated.

1. Fundamental Principles

1.1. Existing Knowledge Base

A considerable body of knowledge has been accumulated regarding the genetic modification and environmental introduction of plants, animals, and microorganisms, and the possible outcomes associated with the introduction of organisms into new ecosystems. Recombinant-DNA techniques have been in use for more than fifteen years in hundreds of laboratories around the world, during which time thousands of different organisms have been modified with these techniques and their characteristics studied. Moreover, much relevant knowledge is available about the great variety of organisms produced by traditional biotechnology methods, such as mutation/selection and cell fusion. This experience provides a sound basis for assessment of possible risks, and for development of additional scientific information where appropriate. Risk assessment related to biotechnology products can and should take advantage of this knowledge base, rather than relying on conjecture.

NATO ASI Series, Vol. G18
Safety Assurance for Environmental Introductions of Genetically-Engineered Organisms
Edited by J. Fiksel and V.T. Covello

1.2. Risk Identification

In principle, the outcomes associated with the introduction into the environment of organisms modified by r-DNA techniques are likely to be the same in kind as those associated with introduction of organisms modified by other methods. Therefore, identification and assessment of the risk of possible adverse outcomes should be based on the nature of the organism and of the environment into which it is introduced, and not on the method (if any) of genetic modification. By the same token, the possible risks associated with the process that produces the organism (e.g. unintended releases from containment) should be distinguished from possible risks arising from the environmental introduction of the organism. To the extent that an organism may have significant novel properties or may be non-indigenous to the environment, risk identification may require special consideration of particular biological or ecological risk mechanisms.

1.3. Safety Assurance

Safety reviews of proposed biotechnology products or applications, such as environmental introductions, have not to date involved formal risk assessments in the strict scientific sense. Risk assessment is generally understood to require the identification of one or more specific undesirable endpoints that might plausibly result from the application (as in environmental risk assessment for carcinogenic pollutants). The initial review and screening procedures currently used for environmental introductions can better be described as "safety assurance", since they have not attempted to quantify or predict any specific endpoints. Only after a plausible risk is identified can a formal risk assessment be performed; the results of such a risk assessment might then be weighed against the anticipated benefits.

1.4. Case-by-Case Approach

The case-by-case approach to safety assurance has been defined by OECD as follows: "Case-by-case means an individual review of a proposal against assessment criteria which are relevant to the particular proposal; this is not intended to imply that every case will require review by a national or other authority since various classes of proposals may be excluded." (OECD, Recombinant DNA Safety Considerations, 1986.) This case-by-case approach, as currently practiced by both governmental and non-governmental bodies, is compatible with the development of generic guidelines and methods for determining when, and against what standards, an application should be screened.

Generic guidelines (such as OECD's GILSP -- Good Industrial Large Scale Practices) help to make the case-by-case approach more efficient, more consistent, and less burdensome, without compromising the safety of performing the research proposed in individual applications. Eventually, with increasing knowledge and experience, it should be possible to reduce futher the number and types of cases that require individual review. For example, categories of exemptions for laboratory and large-scalc work are already part of the NIH RAC Guidelines and the regulations promulgated by various governmental agencies.

2. Methodology Recommendations

2.1. Categorization of Proposed Introductions

To facilitate safety assurance and risk assessment, it is recommended that various categories be defined to reflect the nature of the possible outcomes associated with a proposed introduction and the type of investigation required for risk identification. For example, a review body may wish to expedite or exempt the assessment and testing of well-characterized, low-risk micro-organisms, while taking a more cautious approach with certain classes of higher-risk organisms (e.g., vertebrate pathogens, noxious weeds). For environmental introductions, these categories can be based upon:

- the nature of biological functions(s) affected or introduced
- the environment from which the host organism was taken
- the ecological characteristics of the rDNA organism
- the characteristics of the target environment
- the scale and frequency of the proposed introduction

2.2. Stages of Risk Assessment

In the event that a specific risk has been identified, risk assessment can be facilitated by recognition of several discrete stages in the process whereby introduced organisms can produce various outcomes:

- environmental introduction
- survival
- proliferation
- establishment
- beneficial or adverse effects

Although in certain cases these stages may be tightly interwoven (e.g. proliferation and establishment), it is helpful to identify causal sequences in which the occurrence of later events is dependent upon earlier events. For example, if it can be shown that the probability of establishment in a particular medium (e.g. soil) is negligible, then it may not be necessary to analyze or test adverse effects upon soil ecosystems. This "weakest link" approach, while efficient, may overlook the existence of independent, parallel mechanisms or indirect effect pathways. Moreover, it should be understood that there may always be some, albeit infinitesimal, residual risk, since the risk of any activity can never be reduced to absolute zero.

2.3. Analytical Logic Scheme

Based on the above considerations, some proposed environmental introductions may fall into categories that require detailed risk assessment. In the absence of a uniform safety assurance scheme, most nations have conducted such assessments using expert panels of government regulators, scientists, and representatives of relevant groups. These panels generally follow various "points to consider" guidelines set forth by OECD, EPA, ACGM, NIH, or other groups. The role of the expert group is to evaluate any incremental risks associated with specific genetic modifications of microorganisms intended for proposed beneficial uses in the open environment. The process is often step-wise in that the first stages may include only a preliminary assessment, e.g., to determine if further tests are required to ensure safety for a small scale field trial.

At present, no clear logical scheme has been articulated for how to interpret, evaluate, or critique the information that is requested in the points to consider documents. Each review is dependent upon the judgment of the expert review panel, and there has been no systematic effort to codify the review principles that have been used. It would be desirable to capture the rationales and inferences that have been used in past reviews so that they may be refined and re-applied as necessary.

As a first step towards this goal, and as a means of enhancing the present review mechanism, it is recommended that development of an analytical logic scheme be considered to support and codify the current case-by-case approach. This scheme would be analogous to a "roadmap", in that it would provide the following types of guidance:

- In what order information should be considered
- What questions should be asked at each step of the review
- What criteria should be used to initiate additional investigation

Thus, an analytical logic scheme will go beyond the existing guidelines by suggesting not only "what" to consider but also "whether" and "how" to consider the various points. It will provide an intermediate link between the points to consider documents and rigorous risk assessments for specific areas of concern; in fact, it could be used to determine whether rigorous assessments are appropriate. The expected benefits of this approach are:

- consistency of methodology within and between agencies
- systematic treatment of appropriate beneficial and adverse endpoints
- identification of needs for additional research or testing
- encouragement of scientific consensus across disciplines

- facilitation of external communications about the review process

It is unlikely that a universal logic scheme will be appropriate for the diverse application areas of modern biotechnology. Instead it is recommended that separate schemes be developed to focus upon particular classes of applications (e.g. microbial pesticides).

3. Communication and Cooperation

3.1. International Cooperation

International cooperation is strongly encouraged for all aspects of safety assurance and risk assessment for biotechnology-related applications, including the sharing of scientific information and the development of scientific assumptions and logical review schemes. The existence of international data bases and exchange of information and experience on uses of genetically-modified microorganisms in the environment will markedly enhance scientific assessment capabilities.

Projects such as the joint U.S. Biotechnology Science Coordinating Committee and EEC-Research DG XII effort to establish a database which will include microbial taxonomy, release events, and guidelines are valuable and should be encouraged. Such databases will not only help ensure consistency in decision making by expert committees, but can also assist researchers in companies and universities in determining the most beneficial and safe organisms and risk control procedures for environmental applications in such fields as agriculture, forestry and pollution control.

3.2. Scientific Research and Cooperation

Research in both molecular biology, genetics, and ecology has already contributed an extensive body of information and methods useful for scientific assessment. Closer cooperation between these areas of biology will improve the available knowledge base and enable a more informed assessment of real or conjectural risks. For example, an important research area is the regulation of a metabolic pathway of a cell and its impact on the ecological behavior of the organism. Another important area is the relation between molecular evolution and the evolution, by genetic exchange or mutation, of the ecosystem. The fields of microbial ecology and molecular taxonomy have been under-funded in the past two decades, and should receive increased support.

4. Applied Research Needs

It is recommended that additional support be provided for research on appropriate methods of laboratory testing, microcosm experiments, and contained or open field trials. These methods are often useful in responding to information requirements during safety reviews of proposed environmental introductions. Support should also be provided for those areas of applied research, e.g., microbial marking and environmental detection methods, which have broad applicability to risk management once an organism enters larger scale field trials or commercial use. Such applied research should be carried out in parallel with the more specific regulatory testing, and the interchange of information and data is desirable between those engaged in the two types of efforts.

It is further recommended that funding for applied research related to consequence assessment be made more readily available. The scientific community needs to develop a consensus as to the types of consequences that are possible, and the information needed to assess these consequences. Such research can most usefully be performed as an adjunct to advanced biotechnology research and development projects being carried out by geneticists, molecular biologists and microbiologists. This will help to ensure that areas in which risk assessment is necessary are correctly identified and understood, and that no fictitious risk issues are addressed.

4.1. Public Education

To maintain public confidence in applications of biotechnology it is important that adequate information and education programs be available. In addition to scholastic education, it is important to provide ongoing information about newly evolving possibilities in biotechnology, describing benefits as well as possible problems. Public information must strive to present relevant facts with clarity and precision, and in a manner that is informative to all segments of society.

Workshop Directors
Joseph Fiksel
Vincent Covello

Summary Report Working Group:
Rita Colwell, chair
Peter Lange
Brian Ager
David Shindler

Organizing Committee:
John Cohrssen
Rita Colwell
David Sakura
Curtis Travis

Workshop Participants:
Frank Abbott
Elizabeth Anderson
Patricia Anderson
Metin Bara
Lawrence Barnthouse
Umbertto Bertazzoni
Gustav Brunius
Robert Colwell
Fred Davison
Martin Day
Claire Franklin
Werner Frommer
Helge Gyllenberg
Marcel Ingold
Daphne Kamely
Morten Laake
Ben Lugtenberg
Ernesto Penas
Alan Paton
David Pramer
Parmely Pritchard
Amy Rispin
Bruna Teso
Hisao Uchida
Duncan Veal

P A R T   II

RISK ANALYSIS PERSPECTIVES

"OLD" BIOTECHNOLOGY TO "NEW" BIOTECHNOLOGY: CONTINUUM OR DISJUNCTION?

F.E. Young and H.I. Miller

U.S. Food and Drug Administration

5600 Fishers Lane

Rockville, Maryland 20857

U.S.A.

INTRODUCTION

Since the advent of recombinant DNA techniques and hybridoma technology gave rise more than a decade ago to "new" biotechnology, there has been much discussion of whether the products newly available should be regarded as refinements, or extensions, of previous ones, or as resulting from a technological disjunction. This conundrum is important because of its intimate connection to the question whether existing methods of risk analysis or risk assessment are applicable to or adequate for new processes and products. This is a pivotal issue, because the purpose of such risk assessment is to provide scientists, government regulators and others a measure of the safety attendant to the testing or use of a product, and to provide guidance as to the management of the risk that may be present.

According to Fiksel and Covello[1], the risk assessment literature generally defines "risk" as the potential for adverse consequences of an event or activity. Risk assessment or analysis is the process of obtaining quantitative or qualitative measures of risk levels, including estimates of possible health and other consequences. Implied, of course, are approximations of the uncertainty of those estimates.

NATO ASI Series, Vol. G18
Safety Assurance for Environmental Introductions
of Genetically-Engineered Organisms
Edited by J. Fiksel and V.T. Covello

The components or the techniques of risk assessment may be found elsewhere[2]. We will attempt to place new and old biotechnology in perspective, emphasizing that:

- Biotechnology is not a discrete, unitary entity
- New biotechnology is not as radically novel as is often portrayed
- Vast experience with macroorganisms and microorganisms manipulated by nature or by man provide us an important and useful perspective for current and future applications, both scientific and commercial
- The safety provided by governmental regulation can be bought only at a cost, and there is a point at which more stringent regulation and more expenditure of resources do not confer enhanced safety
- There exist classes of proposed trials -- and certainly individual experiments -- that do not require risk assessments by governmental authorities on each and every proposal

The goal of FDA's approach -- and, indeed that of the entire U.S. government's coordinated framework for the regulation of biotechnology products -- is to limit potential risks, while encouraging the innovation, development and availability of new biotechnology products. We must recognize, however, that not only must products _be_ safe, but the public must have confidence in their safety. In this respect, the advent of new biotechnology poses a major challenge, a challenge that involves perception as much as reality. Only if we can correct misconceptions and eliminate the harmful myths that have grown up around the buzzword "biotechnology" can the potential of the new technology be realized.

THE MYTHS

Myth #1: That Biotechnology is a Discrete Entity

One myth is that biotechnology is something discrete or homogeneous, a corollary of which is that there exists a "biotechnology industry" that can or should be rigidly controlled. This view is facile but inaccurate. Biotechnology is merely a catch-all term for a broad group of useful, enabling technologies with wide and diverse applications in industry and commerce. A useful working definition used by several U.S. government agencies is, "the application of biological systems and organisms to technical and industrial processes." This definition encompasses processes as different as fish farming; forestry; the production of enzymes for laundry detergents; and the genetic engineering of bacteria to clean up oil spills, to kill insect larvae, or to produce insulin. Biotechnology is myriad dissimilar processes producing even greater numbers of dissimilar products for vastly dissimilar applications. Biotechnology processes and products are arguably so diverse and have so little in common with one another that it is difficult to construct valid generalizations about them, for whatever purpose. Putting this another way, biotechnology has no systematic, uniform characteristics that enable it to be legislated or overseen in the homogeneous way that is possible, for example, for the underground coal mining industry.

The diversity of biotechnology has important implications. It dictates that regulation of so many end uses must be accomplished by many government agencies. As the ultimate characteristics and uses of biotechnology's products vary, so does agency jurisdiction over those products (see, for example, references 3-4). And so, of course, does the nature and depth of agencies' evaluation of the products: clearly, EPA's review of an enzyme used as a drain cleaner will be different from FDA's review of the same enzyme injected into patients to dissolve blood clots. The diversity of products and their applications argues against the usefulness of legislation or regulations that attempt to encompass unnatural groupings such as "biotechnology," "genetic engineering," or "deliberate releases"[5].

Myth #2: That Biotechnology and Genetic Engineering are New

A second myth is that biotechnology is new. On the contrary, many forms of biotechnology have been widely used for millennia. Earlier than 6000 B.C., the Sumerians and the Babylonians exploited the ability of yeast to make alcohol and brewed beer. A "picture" of the ancients preparing and fermenting grain and storing the brew has even been retained for posterity in a hieroglyphic[6].

Perhaps the best-publicized subset of biotechnology -- and also one of the oldest -- is "genetic engineering," the manipulation directly or indirectly of an organism's DNA. Genetic engineering dates from man's recognition that animals and crop plants can be selected to enhance desired characteristics. In the traditional, or "conventional," breeding and selection of improved plants, for example, the genetic material of plants is combined to create new and useful traits. In this well-established form of genetic engineering, changes are made at the level of the whole organism; selection is made for desired phenotypes, and the genetic changes, most often poorly characterized, occur comcomitantly. During the past decade or so, new technologies have been developed that enable genetic material to be modified at the cellular and molecular levels, and are more precise and deliberate variants of genetic engineering; the precision of these techniques often provides a better characterized and more predictable product.

Hardy and Glass[7] have carefully distinguished three modes of genetic engineering that constitute a continuum of scientific sophistication and precision: whole organism, cellular, and molecular genetic engineering. They point out that in all three, DNA is modified or combined to increase genetic variation, thereby enlarging the pool of potentially useful traits. The three modes differ not in the end product but in the process used to generate the genetic variability.

In whole organism engineering, or traditional breeding, the genetics of the process is largely random -- entire sets of genes of two animals or plants are combined, with selection for a desired phenotype; concomitant genetic changes are complex and most often poorly-characterized. Despite the relative slowness and laboriousness of this technique, the successes have been monumental, for both interspecific hybridization and intergeneric gene transfer applications. The exploitation of interspecific hybridization for crop improvement is epitomized by the advances in wheat breeding[8]. Gene transfer from related species into cultivated wheat began in 1930, when McFadden reported transferred resistances to stem rust and loose smut diseases from tetraploid emmer wheat to hexaploid bread wheat variety[9]. The resulting bread wheat variety was widely grown in the U.S. and was reponsible for one of the longest rust-free periods in the history of U.S. wheat cultivation. Other genes for resistance to stem rust and powdery mildew and to Hessian fly have since been incorporated into a number of bread wheat varieties. Recent applications of interspecific gene transfer include successful wide hybridization between the cultivated soybean and its wild perennial relatives[10]. Such successful interspecific transfer of traits from wild species to domesticated relatives in the same genus inspired attempts at even more distant crosses, including those between members of different genera. There is certainly evidence that some of our modern crop species, such as rapeseed, tobacco, and wheat, originated in nature by hybridization between different species or genera, and intentional crosses between species in different genera have also successfully transferred specific traits into crop species. Examples include hybridization between cultivated wheat and species of wild grasses from the genera Aegilops, Agropyron, and Secale, in order to transfer traits such as salt tolerance and disease resistance into the crop[11]. There are no known examples of genetically engineered commercial crop species being transformed by these genetic manipulations into troublesome weeds.

The successes of whole organism genetic engineering have not gone unnoticed. The genetic engineering of wheat for human consumption was recognized by the the Nobel Peace Prize in 1970 and the engineering of rice by the Japan Prize in 1987 (ref. 12).

A marked improvement over whole organism genetic engineering, the cellular and molecular technologies provide opportunities for unprecedented precision and specificity, because smaller amounts of better characterized genetic information can be transferred than is possible in a traditional animal or plant cross.

Cellular genetic engineering, which uses the techniques of cell culture and cell fusion to create genetic variation, is less random than engineering at the whole organism level. Thus, fewer variants are necessary to produce organisms with the desired properties, and their selection is easier than at the whole organism level. However, cellular genetic engineering also has limitations: specific genetic changes that result from the engineering may not be known, and there may be changes other than those that confer the desired properties. Cell culture is expected to have a major beneficial effect on agriculture; commercially valuable crops such as sugarcane can be regenerated from cell culture, although this has yet to be achieved for other important crops such as wheat, corn, and soybeans[6].

Molecular genetic engineering employs recombinant DNA and related techniques for genetic constructions. The actual transfer of DNA into plants may be accomplished in several ways: mediated by the Ti plasmid of *Agrobacterium tumefaciens*; direct DNA transfer from culture medium; microinjection into plant protoplasts; or mediated by virus-based gene expression systems.

The molecular techniques make possible a highly specific and precise approach to genetic engineering; the genetic material transferred from one organism to another is usually completely characterized, often consisting of a single gene coding for a hormone or other protein. Molecular genetic engineering has several advantages over whole organism or cellular genetic engineering: specificity and precision of the genetic change; fewer variants created in order to obtain an organism with desired characteristics; the ability to perform experiments in a short time; and the ability to insert synthetic genes (that may not exist at all in nature).

During the past half-century, increased understanding of genetics at the molecular level has added to the sophistication of the genetic engineering of microorganisms. An excellent example is the genetic improvement of *Penicillium chrysogenum*, the mold that produces penicillin: by several methods including screening thousands of isolates and mutagenesis, penicillin yields have been increased more than a hundred-fold during the past several decades. There are many similar examples; microbial fermentation is employed throughout the world to produce a variety of substances including industrial detergents, antibiotics, organic solvents, vitamins, amino acids, polysaccharides, steroids, and vaccines[13]. The value of these and similar products of conventional biotechnology is in excess of $100 billion annually.

In addition to such contained industrial applications, there have been "deliberate releases" of other organisms, including insects, bacteria, and viruses. There are innumerable examples of successful and beneficial "releases," or uses, of live organisms in the environment. Insect release was used successfully to control troublesome weeds in Hawaii early in the 20th century[14]. Other examples are the highly successful program for biological control of St. Johnswort ("Klamath weed") in California by insects in the 1940's and 1950's, and the more recent use of an introduced rust pathogen to control rush skeletonweed in Australia. This remains an area of active research that is promoted by a permanent Working Group on Biological Control of Weeds, overseen jointly by the U.S. Departments of Interior and Agriculture[14].

Currently, more than a dozen microbial pesticidal agents are approved and registered with the U.S. Environmental Protection Agency, and these organisms are marketed in 75 different products for use in agriculture, forestry, and insect control[15]. In another major area, bacterial preparations containing Rhizobium that enhance the growth of leguminous plants (e.g., soybeans, alfalfa, beans) have been sold in this country since the late 19th century; these products allow the plants to produce nitrogen fertilizer from the air.

The most ubiquitous "deliberate releases" of genetically engineered organisms have been during vaccination of human and animal populations with live, attenuated viruses. Live viruses engineered by various techniques and licensed in the U.S. include mumps, measles, rubella, poliovirus, and yellow fever. Inoculation of a live viral vaccine involves not only infection of the immediate recipient, but the possibility of further transmission of the virus and its serial propagation in the community. It is notable that none of the vaccine viruses has become established in the environment, despite their presence there. For example, the presence of vaccine strains of poliovirus in sewage in the U.S. and U.K. reflects the continuing administration and excretion of live virus vaccine rather than its serial propagation in the community[16].

Viral vaccines produced with older genetic engineering techniques have been awesomely effective throughout the world; they are rivaled only by the agricultural "green revolution" as a promoter of human longevity and quality of life. The newest biotechnological techniques, including recombinant DNA, are already providing still more precise, better understood, and more predictable methods for manipulating the genetic material of microorganisms for vaccines.

An analogy is sometimes made between the possible consequences of introducing into the environment organisms manipulated with the new genetic engineering techniques and the ecological disruptions that have been caused by the introduction of certain nonnative (alien or "exotic") organisms; examples cited often include the gypsy moth, the starling, and the kudzu vine. However, this comparison is specious, depending largely on the assumption that rDNA manipulations can alter the properties of an organism in a wholly unpredictable way that will cause it to affect the environment adversely. As discussed below, both theory and experience indicate that this is very unlikely. Genetically manipulated organisms, whether engineered by conventional or new molecular techniques, closely resemble the parent organism, and are, in fact, often at an evolutionary disadvantage with respect to their parents and cohorts.

Myth #3: "The unknowns far outweigh the knowns where the ecological properties of microbes are concerned."[17]

This is an excessively negative statement, and is particularly dubious for many microorganisms of commercial interest, including Pseudomonas syringae, Thiobacillus species, Bacillus thuringiensis, Bacillus subtilis, Rhizobium, and Baculovirus, to name a few. Also, it should be noted that many microbes are essential to ecosystem processes or otherwise beneficial to man, and that only a minuscule fraction of microbes are pathogenic or otherwise harmful.

In the heading above, one could just as easily substitute "the functions of the mutations in polio virus vaccine" for "the ecological properties of microbes;" these unknowns have not prevented our using live, attenuated polio virus vaccine safely and effectively for three decades. Similarly, they have not prevented the unregulated small-scale testing of innumerable different microbes in the environment -- small-scale field trials were exempt from both the U.S. pesticide and toxic substances statutes until recently -- and which boasts an admirable safety record. The scientific method and prior experience applied logically to risk assessment do enable us to make useful predictions.

Myth #4: That New Genetic Engineering Techniques Will Create Novel,

Dangerous Organisms

The degree of novelty of microorganisms or macroorganisms created by the new genetic engineering techniques has been widely exaggerated. A corn plant that has incorporated the gene for and synthesizes the *Bacillus thuringiensis* toxin is still, after all, a corn plant. *E. coli* K-12 that has been programmed to synthesize human interferon alpha by means of recombinant DNA techniques really differs very little from its unmanipulated siblings that manufacture only bacterial molecules. Moreover, nature has already tried out innumerable recombinations between even very distantly-related organisms, via several mechanisms (see, for example, ref. 17). Bacteria in nature have long been exposed to DNA from lysed mammalian cells -- for example, in the gut, in decomposing corpses, and in infected wounds. The human population alone excretes on the order of 10^{22} bacteria per day; hence over the past 10^{6} years, many mammalian-bacterial hybrids are likely to have appeared and been tested by natural selection. An analogous argument can, of course, be made for recombination among fungi, bacteria, viruses, and plants. Kilbourne[16] has emphasized that both genetic and ecological constraints operate to prevent the emergence of hyper-virulent viral variants, even though single point mutations can alter virulence. And while nature does occasionally produce, on a time scale that we can observe, a modified pathogen (such as an influenza virus with increased virulence, or HIV-1), we must ask how likely it is that it would do so in one fell swoop from a non-pathogen [*vide infra*]; the chances of such an event arising from the small-scale man-made changes must be compared with the tremendous background "noise" in nature.

Myth #5: That Genetic Manipulation Will Transform a Non-pathogen into a Pathogen

An often-mentioned concern is that genetic manipulation may inadvertently transform a non-pathogen into a pathogen. However, this view ignores the complexity and the multi-factorial nature of pathogenicity. Pathogenicity is not a trait produced by some single omnipotent gene; rather, it requires the evolution of a special set of properties that involve a number of genes. A pathogen must possess two general characteristics, which are themselves multi-factorial. First it must be able to metabolize and multiply in or upon host tissues; that is, the oxygen tension and pH must be satisfactory, the temperature suitable, and a favorable nutritional milieu available. Second, assuming an acceptable range for all of the many conditions necessary for metabolism and multiplication, the pathogen must be able to resist host defense mechanisms for a period sufficient to reach the numbers required actually to produce disease. Thus, the organism must be meticulously adapted to its pathogenic life-style, and even a gene specifying a potent toxin will not convert a harmless bacterium into an effective source of epidemics -- or even localized disease -- unless many other required traits are present. These include, at the least, resistance to host defenses, ability to adhere to specific surfaces, and the ability to thrive on available nutrients provided by the host. And although no one of these traits confers pathogenicity, a mutation that affects any essential one can eliminate it. Moreover, severe pathogenicity is more demanding, and much more rare in nature than mild degreees of pathogenicity, and so the probability of inadvertently creating an organism capable of a medical catastrophe must be vanishingly small.

Myth #6: That All Technology is Intrinsically Dangerous

Another myth is that the application of all new technology is dangerous. The bases of this may be atavistic fears of disturbing the natural order and of breaking primitive taboos, combined with the complexity for the non-scientist of the statistical aspects of risk. The promulgators of this myth who seek to discredit biotechnology eagerly cite the hazards of toxic chemical waste dumps and the technical problems of the nuclear industry, but conveniently ignore the overwhelming successes of telephonic communication, vaccination, blood transfusions, microchip circuitry, and the domestication of animals, plants, and microbes. It is worthy of note that some predicted electrocution from the first telephones, the creation of human monsters by Jenner's early attempts at smallpox vaccination, and the impossibility of matching blood for transfusions. They said, in effect, "The costs will be too high; there is no such thing as a free lunch."

No reponsible person would suggest that some novel practices, processes or products of biotechnology could not be hazardous in some way. Some of these are already well known: workers purifying antibiotics have experienced allergic reactions; beekeepers have been stung; laboratory workers have inadvertently sucked up bacteria through a pipette and suffered gastroenteritis or worse; and vaccines occasionally elicit adverse reactions.

In addition, there have been examples of introductions of exotic species, such as English sparrows and gypsy moths, that have had serious economic consequences. However, the introduction of exotic species is not a useful model for the kinds of organisms being contemplated with new biotechnology. Generally, introductions will be of <u>indigenous</u> organisms that differ minimally (often by only the insertion or deletion of a single structural gene) from organisms already present in the environment, and that will not enjoy a selective advantage over their wild-type cohorts. They will be subject to the same physical and biological limitations of their environments as their unmodified parents. As noted above, a corn plant that has incorporated the gene for and synthesises the Bacillus thuringiensis toxin is, after all, still a corn plant. Thus, a more applicable model for organisms modified by the techniques of new biotechnology is the selective breeding and testing of domesticated plants, animals and microbes, with which there is vast experience and which boasts an admirable safety record.

In any case, a complex and comprehensive regulatory apparatus based in numerous federal agencies in the U.S. has long overseen the safety of food plants and animals, pharmaceuticals, pesticides and other products that can be produced by biotechnology. This should continue to be equal to the task, and to perform in a way that does not stifle innovation. There may never be a "free lunch," but often we can make it an excellent value.

THE APPLICABILITY OF RISK-ASSESSMENT METHODS FOR ENVIRONMENTAL APPLICATIONS OF BIOTECHNOLOGY

Among those scientifically knowledgable about the new methods of genetic manipulation, there is wide consensus that existing risk assessment methods are suitable and applicable for environmental applications of new biotechnology. Several appropriate risk assessment alternatives are available, including: deterministic consequence analysis with confidence bounds; qualitative screening; and probabilistic risk assessment[2]. While it is true that risk assessment is not an exact, quantitative, and predictive discipline, we agree with a National Science Foundation report's conclusion that available methods provide a useful foundation and "a systematic means of organizing a variety of relevant knowledge about the behavior of microorganisms in the environment"[2]. The need for risk assessment of new biotechnology is not new methods, but rather in ascertaining the correct underlying assumptions. For risk assessment as for many other aspects of new biotechnology, new products manufactured with new processes do not necessarily require new regulatory or scientific paradigms.

The above view is supported by the recent report of the U.S. National Academy of Sciences, "Introduction of Recombinant DNA-Engineered Organisms into the Environment: Key Issues[19]." This landmark report has wide-ranging implications in the international community, by providing an authoritative perspective on planned introductions. Several of the most significant of its conclusions and recommendations are:

- R-DNA techniques constitute a powerful and safe new means for the modification of organisms;

- Genetically modified organisms will contribute substantially to improved health care, agricultural efficiency, and the amelioration of many pressing environmental problems that have resulted from the extensive reliance on chemicals in both agriculture and industry;

- There is no evidence that unique hazards exist either in the use of rDNA techniques or in the movement of genes between unrelated organisms;

- The risks associated with the introduction of rDNA-engineered organisms are the same in kind as those associated with the introduction of unmodified organisms and organisms modified by other methods; and

- The assessment of risks associated with introducing rDNA organisms into the environment should be based on the nature of the organism; based on the environment into which the organism is to be introduced;

and independent of the method of engineering *per se*.

The conclusions and recommendations of the National Academy of Sciences report have been echoed elsewhere. Examples include the recent deliberations of an OECD Working Group on Safety of Biotechnology; and the results of a conference held in Bellagio in September 1987 ("Introduction of Genetically-Modified Organisms Into the Environment: A Statement from the Scientific Committee on Problems of the Environment (SCOPE) and the Committee on Genetic Experimentation (COGENE)").

We can summarize the current situation regarding the regulation of new genetic engineering products in a syllogism. Industry, government, and the public already have considerable experience with "deliberate release" of traditional genetically engineered products, including *Rhizobia* for agriculture and live vaccines such as measles and polio. Existing regulatory schemes have protected human health and the environment while simultaneously stimulating industrial innovation. As noted above, there is no evidence that unique hazards exist either in the use of rDNA techniques or in the movement of genes between unrelated organisms. *Therefore, there is no need for additional regulatory mechanisms to be superimposed on pre-recombinant-DNA regulation*.

IMPLICATIONS FOR GOVERNMENTAL POLICY-MAKERS

Of concern to many practitioners and regulators of biotechnology are two non-scientific issues that could play a dominant role in the future of biotechnology development and use in the U.S. and abroad. First, the regulatory climate could, if not rationalized or if inappropriately risk-averse, impede industry's eventual introduction of products being developed now. This could lead to future withdrawal or diminution of new product development, with continued reliance on less sophisticated -- and often more hazardous -- alternative technologies. Second, concerns within the financial community about the long term stability and success of companies doing business in environmentally regulated fields could depress company values, drying up capital for the continued development and testing necessary to satisfy regulatory requirements[19].

Those who are associated with the regulatory issues of biotechnology have a critical responsibility to act quickly and definitively to balance the various opposing forces facing this technology. We reject the notion that all products derived from new biotechnology defy useful, accurate risk assessments or are too dangerous to introduce into the environment: both theory and experience repudiate this assertion. We must be guided in part by the knowledge that there are genuine costs of overly risk-averse regulatory policies that prevent the testing and approval of new products; crops destroyed by frost and the continued application of dangerous chemical pesticides while the "ice-minus" bacteria and new bio-rational pesticides languish untested for years represent a significant toll. At the same time, we must acknowledge legitimate concerns about the safety of product testing and use. The principles that govern the safe use of these products and allow the underlying research to proceed must evolve and be refined. In such a fast-moving technological environment, it is necessary to reappraise regularly the scientific basis of existing regulation and to make any required adjustments in either the technology of regulation or the statutory basis for regulation.

The professional practitioners, regulators and observers of biotechnology must strive to demystify it and to provide the proper perspective for the public, because it is the public who will benefit most. The stakes are high both in economic terms and in terms of social benefit. In the past year, the U.S. Food and Drug Administration has approved several products of new biotechnology that are medical milestones. These include alpha-interferons for the treatment of a lethal leukemia, a monoclonal antibody preparation for preventing rejection of kidney transplants, and a new-generation hepatitis vaccine. Among myriad other applications, biotechnology promises vaccines against scourges such as malaria, schistosomiasis, and AIDS, and new therapies that could ameliorate or cure for the first time such genetic diseases as sickle-cell anemia or certain inherited immune deficiencies. Used for new generations of medicines and food plants and animals, it could provide partial solutions to the trinity of despair -- hunger, disease, and the progressively deteriorating mismatch between material resources and population.

SUMMARY

Biotechnology and its subset, genetic engineering, have been widely applied for millenia, including innumerable successful and beneficial uses, or "releases," in the environment. The precision and power of genetic manipulation of both macroorganisms and microorganisms have increased during the past half century with increased understanding of molecular genetics. The techniques of "new biotechnology" are generally viewed in the U.S. as extensions -- refinements -- of older techniques for genetic manipulation. For these reasons, among those scientifically knowledgable about the new methods of genetic manipulation, there exists wide consensus that current risk assessment methods are suitable and applicable for environmental applications of new biotechnology; new products manufactured with new processes do not necessarily require new regulatory paradigms. Finally, there are genuine costs of overly risk-averse regulatory policies that prevent the testing and approval of new products; such policies are anti-innovative, anti-competitive, and delay the benefits of the new products to the public. These policies both emanate from and feed a pernicious anti-science movement that threatens basic scientific research as well. The long term remedy must be improved public education about science and technology to produce a generation with at least enough knowledge to avoid being "bamboozled by foolishness."[20]

REFERENCES

1. Fiksel, J.R. and V.T. Covello, "An Overview and Evaluation of the Suitability and Applicability of Risk Assessment Methods for Environmental Applications of Biotechnology," in, "The Suitability and Applicability of Risk Assessment Methods for Environmental Applications of Biotechnology," V.T. Covello and J.R. Fiksel, eds., U.S. National Science Foundation, Washington, D.C., 1985.

2. The Suitability and Applicability of Risk Assessment Methods for Environmental Applications of Biotechnology, V.T. Covello and J.R. Fiksel, eds., U.S. National Science Foundation, Washington, D.C., 1985.

3. Federal Register 49, 50856 (1984).

4. Federal Register 51, 23302 (1986).

5. Miller, H.I., Pharmaceutical Engineering 6, 28 (1986).

6. Demain, A.L. and N.A. Solomon, Sci. Amer. 245, 66 (1981).

7. Hardy, R.W.F. and D.J. Glass, Issues in Science and Technology 1, (1985).

8. Goodman, R.M., H. Hauptli, A. Crossway, and V.C. Knauf, Science 236, 48 (1987).

9. McFadden, E.S., J. Am. Soc. Agron. 22, 1050 (1930).

10. Newell, C.A. and R. Hymowitz, Crop Sci. 22, 1062 (1982).

11. Rick, C.M., J.W. DeVerna, R.T. Chetelat, M.A. Stevens, Proc. Natl. Acad. Sci. U.S.A. 83, 3580 (1986).

12. Miller, H.I. and F.E. Young, JAMA 257, 2334 (1987).

13. Health Impact of Biotechnology: Report of a WHO Working Group, Swiss Biotech. 2, 7 (1985).

14. Klingman, D.L. and J.R. Coulson, Plant Dis. 66, 1205 (1982).

15. Betz, F., M. Levin, and M. Rogul, Recomb. DNA Tech. Bull. 6, 135 (1983).

16. Kilbourne, E.D., Epidemiology of Viruses Genetically Altered by Man -- Predictive Principles, in Banbury Report 22, Genetically Altered Viruses and the Environment, Cold Spring Harbor Laboratory, 1985.

17. Sharples, F.E., Science 235, 1329 (1987).

18. Davis, B.D., Science 193, 442 (1976).

19. Kingsbury, D.T., Bio/Technology 4, 1071 (1986).

20. Kennedy, D., The Wall Street Journal, p. 11, October 29, 1987.

Potential Applications of Knowledge System Technology to Biotechnology Safety Assurance

Joseph Fiksel, Ph.D.
Principal Scientist
Risk and Decision Systems
Teknowledge, Inc.
P.O. Box 10119
Palo Alto, CA 94303

1. Introduction

The rapid advance of modern biotechnology from a laboratory science to an emerging industry has been accompanied by considerable debate over appropriate means for regulating biotechnology products. One of the more controversial areas has been the planned introduction of genetically-engineered microorganisms into the environment, for example as an agricultural aid. A central issue in this debate is whether we possess appropriate methods for assessing the hypothetical human or ecological risks associated with such planned releases.

Most previous risk assessment methods have addressed either equipment failures, which could be mathematically simulated, or biochemical agents, for which statistical estimates of dose-response relationships could be developed. Microorganisms released into an ecosystem may not be easily amenable to these types of quantitative analyses, because their characteristics are only partially understood, and because they may proliferate and adapt to the environment in unexpected ways. Perhaps the best analogue is the introduction of a new vaccine, which requires successively more refined tests to discover any adverse consequences.

In a U.S. study sponsored by the White House Office of Science and Technology Policy and the National Science Foundation [1], available risk assessment methods were evaluated for their applicability to environmental introductions of genetically-engineered microorganisms. This study concluded that:

> "Existing scientific knowledge and methods are adequate to perform qualitative screening of specific environmental applications using modified microorganisms developed from organisms with well-defined characteristics.

NATO ASI Series, Vol. G18
Safety Assurance for Environmental Introductions
of Genetically-Engineered Organisms
Edited by J. Fiksel and V.T. Covello

> Further knowledge is necessary to advance risk assessment methods to the point where quantitative, predictive analysis can be performed for a range of modified microorganisms developed from microorganisms with poorly-defined characteristics." (p. 31)

The study suggested that, given the state of the art of risk assessment, the preferred approach was to apply qualitative screening methods backed up by testing and monitoring procedures as deemed necessary. In fact, this type of approach is already being practiced in most countries.

A recent international workshop on the same subject [2], observed that safety reviews of proposed such as environmental introductions do not involve a formal risk assessment in the strict scientific sense. Risk assessment is generally understood to require the identification of one or more adverse endpoints that might result from the application (as in environmental risk assessment for carcinogenic pollutants). The initial review and screening procedure for environmental introductions can better be described as "safety assurance," since there may not be any specific identifiable risks. Only after a possible risk is identified can a formal risk assessment be performed; the results of such a risk assessment might then be weighed against the anticipated benefits.

Thus, an analogy may be drawn between safety assurance and the quality assurance practices that are routinely applied in manufacturing. One can view the screening activities of regulatory bodies as testing the quality of a proposed introduction to ensure that it satisfies certain requirements and standards. Risk assessment is then analogous to exploring the potential consequences of a possible defect, while risk management is analogous to reducing the chances of a significant defect occurring. The risk assessment and management process is discussed further in Section 2 below.

Given the above perspective, this paper suggests that there are several critical requirements for effective safety assurance in the context of environmental introductions of genetically-engineered microorganisms:

1. availability of a comprehensive scientific and regulatory knowledge base regarding past introductions and current policies;

2. a methodology for accessing and interpreting that knowledge, and systematically applying the relevant policies and criteria.

Thanks to recent advances in knowledge system technology, as described in Section 3, it appears that the use of computer-based systems is a feasible and attractive means of meeting these requirements. Section 4 suggests that knowledge systems can provide reliable and consistent support to humans in the difficult task of evaluating proposed introductions, and particularly in the qualitative screening exercise that is typically the first stage of review. This notion is elaborated and illustrated in Sections 5 and 6.

2. Qualitative Methods in Risk Assessment and Management

There have been numerous efforts over the past few years to make a distinction between "risk assessment" and "risk management". The National Academy of Sciences defined risk assessment as the use of scientific methods, models and data to develop information about specified risks, and risk management as the subsequent balancing of risks against other criteria in order to make decisions about risk mitigation or control [3]. More recently, a National Science Foundation study suggested a three-stage model of risk assessment, risk evaluation, and risk management, with the middle stage focusing on analysis of risk/benefit tradeoffs prior to decision-making [4].

Regardless of the model chosen, there is general agreement that the initial stages of risk assessment are largely objective and scientifically-based, whereas the latter stages of risk management inevitably incorporate policies and value judgments based on economic, political, and social factors. It has been argued, therefore, that risk assessment and risk management should be kept separate, to avoid the premature introduction of values into the science of risk assessment. In practice, however, judgmental factors inevitably influence the risk assessment process [5].

There are three major categories of risk assessment that can be distinguished, with different levels of precision and rigor. The first category is "empirical" risk assessment, which is based on scientific evidence and real-world experience. For example, the science of epidemiology relies upon empirical information to draw statistical conclusions about the potential for disease in populations exposed to various environmental risk factors.

The second category is "model-based" risk assessment, which uses predictive models in place of empirical statistical data. For example, when a company introduces a new chemical product, they perform a series of toxicological tests which are used in regulatory submissions as evidence that the product is safe. Since the product cannot be tested on human populations, scientists must predict its degree of risk using toxicological models for dose-response extrapolation or physical models for absorption of a material into the body. Use of these models usually requires scientific assumptions, and results in a simplified representation of reality which introduces uncertainty into the risk assessment process. Nevertheless, model-based risk assessments are regularly used by regulatory agencies such as the U.S. Environmental Protection Agency to support decision-making.

Finally, the third category is "qualitative" risk assessment, which involves the use of heuristics and logical schemes to derive plausible conclusions. Generally, qualitative risk assessment is used when there are neither sufficient empirical data nor reliable predictive models available, or when available resources are limited. Instead, qualitative methods rely upon good scientific judgement to assess the combinations of factors that might contribute to a risk. This method of risk assessment can be just as useful as the other two, even though it does not strive for mathematical precision.

In the biomedical arena, for example in the field of medical device regulation, qualitative assessment is often used for decision-making by the U. S. Food and Drug Administration. Rather than trying to predict the chances of adverse outcomes, the agency applies a stringent set of qualitative screening criteria to assure that a product is as safe as can be expected under the circumstances, and therefore is acceptable for use. While it foregoes numerical predictions, this type of methodology has proven very useful in practice.

For many innovative types of environmental introductions, empirical risk assessment methods will be impractical due to the lack of an adequate data base of organism characteristics and interactions with the target ecosystem. Similarly, model-based assessment will be difficult or impossible due to the current lack of adequate predictive models of environmental outcomes for specific ecosystems. Therefore, for the foreseeable future, it is likely that biotechnology risk assessment will rely heavily upon qualitative screening methods such as those currently practiced by the National Institutes of Health Recombinant DNA Advisory Committee.

While the qualitative approach appears to be sensible and appropriate, it does have certain potential limitations. Assessments are generally carried out using expert panels of government regulators, scientists, and representatives of relevant groups, who follow various "points to consider" guidelines set forth by OECD, EPA, ACGM, NIH, and other groups. However, at present, no clear logical scheme has been articulated for how to interpret, evaluate, or critique the information that is requested in the points to consider documents. Each review is dependent upon the judgment of the expert review panel, and there have been few systematic efforts to codify the underlying review principles. It would be desirable to capture the rationales and inferences that have been used in past reviews so that they may be refined and re-applied as necessary. It is argued below that knowledge system technology can help to provide this type of capability.

3. Knowledge System Technology

The term "artificial intelligence" (AI) often conjures up images of industrial robots and language translation machines. However, one of the fastest-growing application areas for AI is in management decision-making support for both business and government. In particular, AI promises to have a significant impact on professionals working in the field of health, safety, and environmental risk management [5].

AI scientists have created a new class of computer software, called "knowledge systems," which can simulate human reasoning [6]. Knowledge systems are usually defined as computer programs that give advice, solve problems, or perform other "intelligent" tasks. More generally, knowledge systems automate the application of knowledge, just as calculators automate the application of arithmetic or data-base management programs automate the indexing and retrieval of records. Thus, knowledge systems technology involves the computer-aided representation of human knowledge and reasoning in symbolic form. The design and development of knowledge systems requires a specific type of expertise called "knowledge engineering" [7].

The potential uses of knowledge systems are only beginning to be explored. Early applications tended to view the knowledge system as an independent agent that offered expertise in response to user requests. In fact, knowledge systems can form the basis of a much broader class of support tools that extend the scope and power of human reasoning. In the near future, risk analysts and risk managers will be assisted in their day-to-day activities by powerful workstations that simplify much of the effort involved in searching for information, running analytic models, and interpreting the results [8].

At the present time, the most common risk-related applications of knowledge system technology involve the use of "expert systems," which provide specialized technical advice based on knowledge extracted from human experts. For example, a number of expert systems have been developed that advise emergency response teams about how to deal with industrial accidents such as chemical spills. In the regulatory arena, the Environmental Protection Agency is exploring the use of expert systems to assist in hazardous waste site permitting, in water quality modelling, and in a number of other environmental engineering applications [9]. Many large manufacturing companies have established internal AI groups that are building expert systems to advise on diagnosis and repair of equipment failures, product safety testing, and a host of similar tasks. This rapid expansion of expert system applications has been made possible by the commercial availability of low-priced general-purpose development tools, called "shells," which have shortened the typical system development time by an order of magnitude, from years to months.

However, these early achievements are only the beginning of what promises to be a profound change in the way risk analysts and managers utilize computer software technology. Recent innovations have made it possible for knowledge system technology to provide a decision support environment, in which a broad range of knowledge is captured and applied to a variety of ongoing risk management activities. System users can simultaneously access qualitative knowledge bases, quantitative data bases, conventional analytic models, and expert advisory systems. As demonstrated in Section 6 below, the necessary technology to develop these systems already exists, and numerous prototypes of "intelligent decision support systems" have been built.

The following section points out a number of specific areas in which these types of intelligent systems can support safety assurance for environmental introductions.

4. Principles for Biotechnology Safety Assurance

An international group of 35 scientific and regulatory experts participated in a NATO Advanced Research Workshop in Rome, Italy in June of 1987. The objectives of the workshop were to evaluate the available safety assurance and risk assessment methods for environmental introductions of genetically-engineered organisms and to establish principles for improving these methods [2]. Below are summarized the major principles that emerged from this workshop and the potential contributions of knowledge system technology in implementing these principles.

Principle 1: Use of Existing Knowledge Base

> "A considerable body of knowledge has been accumulated regarding the genetic modification and environmental introduction of plants, animals, and microorganisms, and the possible outcomes associated with the introduction of organisms into new ecosystems ... Risk assessment related to biotechnology products can and should take advantage of this knowledge base, rather than relying on conjecture."

Knowledge systems permit the electronic capture, maintenance, and distribution of a large body of knowledge regarding technical, economic, or scientific aspects of various risks. Establishment of a single conceptual framework for accumulation of knowledge from multiple sources can help to clarify the nature of scientific controversies and disagreements about facts and assumptions, and can provide a basis for continual improvement of the knowledge base in the light of new information.

Principle 2: Risk Identification Criteria

> "Identification and assessment of the risk of possible adverse outcomes should be based on the nature of the organism and of the environment into which it is introduced, and not on the method (if any) of genetic modification ... To the extent that an organism may have significant novel properties or may be non-indigenous to the environment, risk identification may require special consideration of particular pathways or mechanisms of risks."

Knowledge systems can assist in screening proposed environmental introductions for characteristics that might conceivably lead to adverse outcomes. They can apply both quantitative and qualitative screening criteria, and can recommend followup actions or priorities depending upon the results. Thus, use of knowledge systems as a preliminary review mechanism can assure consistent and systematic application of relevant criteria.

Principle 3: Use of Generic Guidelines

> "The case-by-case approach to safety assurance is compatible with the development of generic guidelines and methods for determining when, and against what standards, an application should be screened...generic guidelines help to make the case-by-case approach more efficient, more consistent, and less burdensome, without compromising the safety of individual applications."

Knowledge systems can help to implement generic guidelines by systematically and exhaustively reviewing all proposed introductions aginst each relevant guideline. Thus, knowledge systems can expedite the often-burdensome process of determining the applicable health, safety, and environmental regulations, and ensuring that all requirements have been followed.

Principle 4: Categorization of Proposed Introductions

> "To facilitate safety assurance and risk assessment, it is recommended that various categories be defined to reflect the nature of the possible outcomes associated with a proposed introduction and the type of investigation required for risk identification...these categories can be based upon the nature of biological functions(s) affected or introduced, the environment from which the host organism was taken, the ecological characteristics of the rDNA organism, the characteristics of the target environment, and the the scale and frequency of the proposed introduction."

Knowledge systems are well-suited to the definition and application of qualitative classification schemes. Once the categories have been established, a knowledge system can easily recognize the appropriate classification for a proposed introduction and initiate whatever review procedures are required. For example, a specific category might trigger particular risk identification criteria or generic guidelines, as discussed above.

Principle 5: Analytical Logic Scheme

> "It is recommended that development of an analytical logic scheme be considered to support and codify the current case-by-case approach. This scheme would suggest in what order information should be considered, what questions should be asked at each step of the review, and what criteria should be used to initiate additional investigation. Thus, an analytical logic scheme will go beyond the existing guidelines by suggesting not only 'what' to consider but also 'whether' and 'how' to consider the various points."

Knowledge systems are capable of representing and executing complex logical schemes involving a specific domain of investigation. Most knowledge systems will seek the necessary information to pursue a logic scheme from the current knowledge base, and if it is not available they will request additional knowledge from the system user. They are also capable of working with incomplete or missing information, and can interact effectively with the user in selecting alternative strategies.

The NATO workshop concluded that the use of an analytic logic scheme offers a number of potential advantages to those involved in the safety assurance process:

- consistency of methodology within and between agencies
- systematic treatment of possible beneficial and adverse endpoints
- identification of needs for additional research or testing
- encouragement of scientific consensus across disciplines
- facilitation of external communications about the review process
- accumulation of an organized knowledge base for future reference

The following section describes in greater detail how such logic schemes might be constructed using knowledge system technology.

5. Design of Analytical Logic Schemes

The basic methodology of knowledge engineering can be applied to the design of logic schemes that help to carry out the customary steps in the risk assessment and management process. From a knowledge engineering point of view, there are several types of human knowledge that influence safety assurance decisions in biotechnology applications:

- empirical knowledge about organisms and their environment
- situational knowledge about test conditions and objectives
- judgmental knowledge about human beliefs and priorities
- theoretical knowledge about ecological relationships
- normative knowledge about policies and acceptance criteria

A knowledge system that logically introduces these types of knowledge can be useful throughout the various stages of risk assessment and risk management. Figure 1 depicts the cumulative application of knowledge, with the possibility of terminating at each stage. Any analytical logic scheme that supports safety assurance will probably be structured according to this general framework. The individual stages are discussed briefly below.

Risk identification is the first step of the process, and involves applying screening criteria and logical deduction to the proposed introduction that is being considered. For example one possible risk identification criterion might be:

The organism was modified with genetic material from a pathogen.

If no possible adverse outcomes are detected in this risk identification step, then there may be no need for further screening. If certain classes of potential risk are identified, then the analytical logic scheme can invoke screening criteria that are relevant to each class. For example, a judgmental screening rule relevant to pathogenic risks might be:

If pathogenic traits are present in the donor or recipient organism
  and intergeneric transfer of genetic material is plausible
then a high level of concern should be assigned to this proposal.

At this point, specific risks (i.e., endpoints) might be identified and assigned a sufficient level of concern to warrant further assessment. The analytical logic scheme may suggest that certain scientific models be invoked, such as event tree analysis models, that can produce theoretical estimates of the degree of potential hazard. For example, a model might compute the expected likelihood of an adverse impact upon indigenous microorganisms using the following simple formula:

$$\text{Probability of impact} = P1*P2*P3*P4$$

where P1 = prob. of survival
P2 = prob. of dissemination beyond test plot
P3 = prob. of non-target plant colonization
P4 = prob. of competitive advantage

Naturally, such estimates are based on numerous assumptions and may have large associated uncertainties. Biotechnology risk assessment methods are relatively immature, and in many cases there will be insufficient data to support quantitative analysis. Nevertheless, an analytical logic scheme can instead use qualitative models to establish the nature, if not the degree, of possible risks. For example, rule-based reasoning might be used to derive predictions as follows:

if the microorganism survives
  and disseminates beyond the test plot
  and colonizes non-target plants
  and has a competitive advantage
    then there is a moderate likelihood that
      it may displace indigenous species

Additional rules can be developed to investigate each of the premises in the above rule, thus providing a chain of inferences. This is precisely how rule-based knowledge

Figure 1

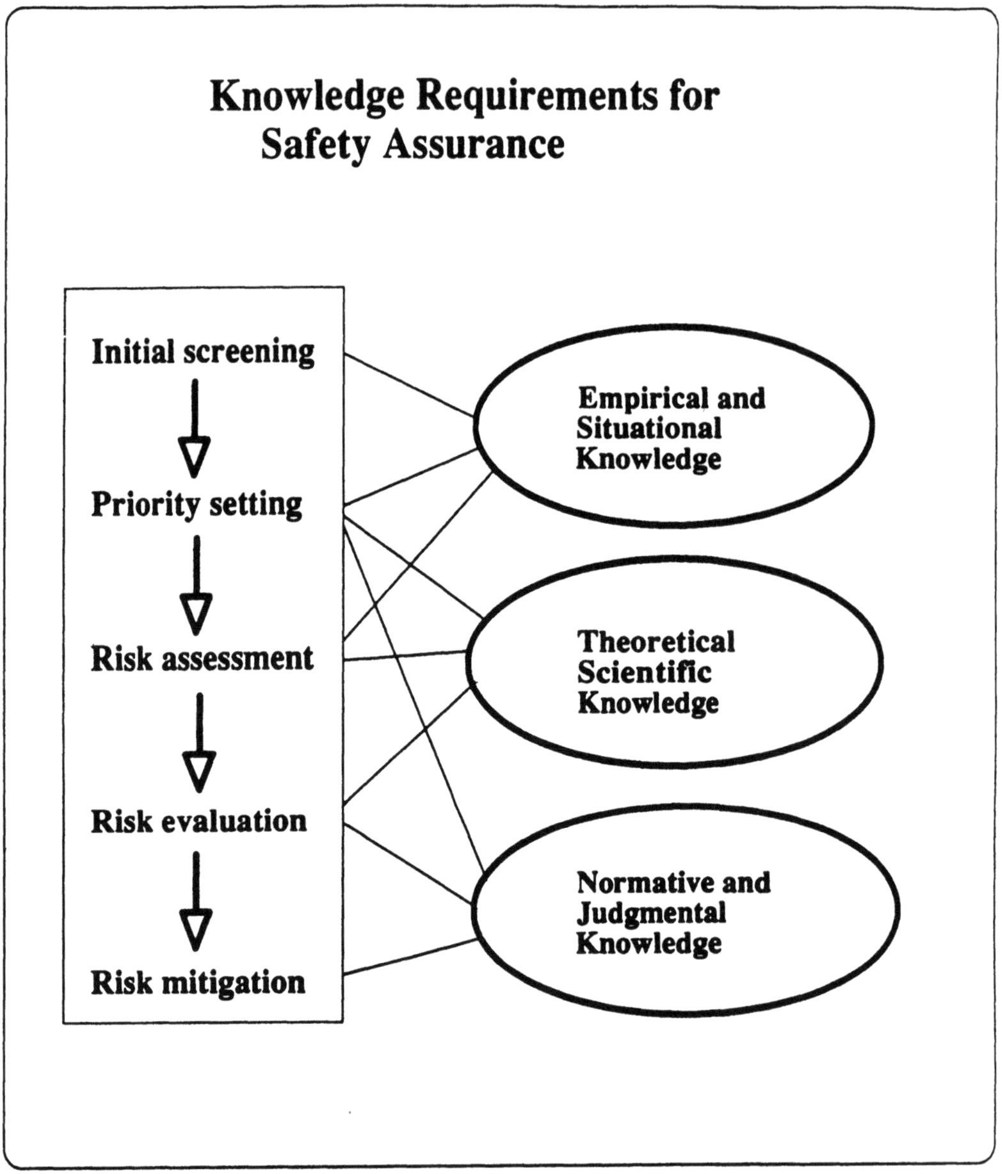
Knowledge Requirements for Safety Assurance
Initial screening
Priority setting
Risk assessment
Risk evaluation
Risk mitigation
Empirical and Situational Knowledge
Theoretical Scientific Knowledge
Normative and Judgmental Knowledge

systems are structured.

Once potential risks have been assessed, it is possible to introduce value judgments regarding the degree of concern about a specific hypothesized endpoint. An analytical logic scheme can assist in this evaluation by providing comparisons with other similar applications. The final step in the process requires normative judgments about appropriate methods for controlling or reducing risks. Choice of these methods may be based upon logical rules, or may require more formal risk/benefit balancing methods. While knowledge system support is certainly appropriate for these latter stages of risk management, it is beyond the scope of this paper.

Figure 2

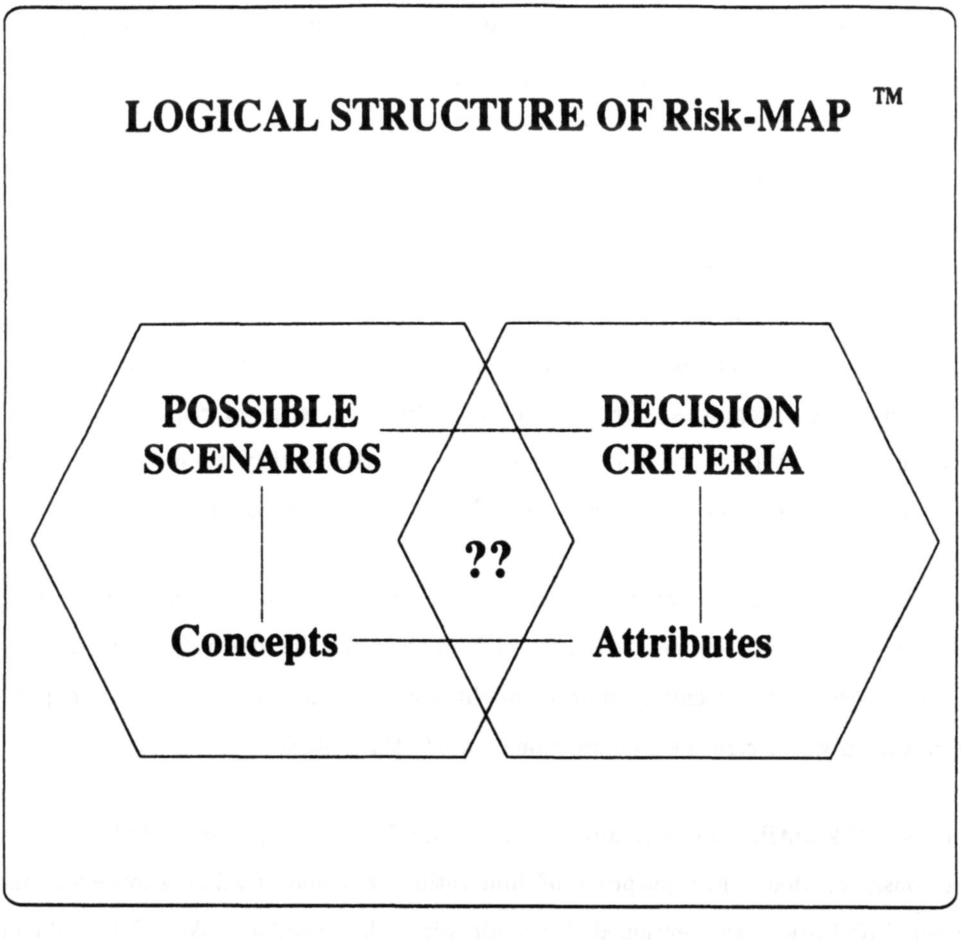

6. Example of a System for Safety Assurance

6.1. General Description of Risk-MAP

In order to demonstrate the type of analytical logic scheme described above, we have used Teknowledge's Risk-MAP (tm) technology to construct a simple protoype system that assists with qualitative and quantitative decision-making in the safety assurance process. Risk-MAP is an intelligent decision support system that helps to organize and apply human knowledge for purposes of managing risks. It is implemented in M.1, Teknowledge's patented expert system development package; therefore it can run on an IBM PC AT or compatible personal computer.

Risk-MAP is designed as a day-to-day decision support tool for professional analysts and decision-makers in a wide variety of fields. It allows users to maintain and explore knowledge about their sphere of interest, including possible risk scenarios and decision criteria. Then, through systematic logical reasoning, it can screen, evaluate, or compare alternative scenarios and recommend action priorities.

A central notion in Risk-MAP is the use of "scenarios" to describe existing or hypothetical situations. As illustrated in Figure 2, scenarios are composed of "concepts" which represent aspects of the situation. Concepts, in turn have "attributes," or properties. Criteria can be defined in terms of required attributes, and then scenarios can be tested to see whether they satisfy these criteria. This knowledge representation approach is described further in Section 6.2.

Example: A scenario about environmental introduction of microorganisms might include the concept "soil ecosystem," which might have the attribute "high diversity of microflora". A screening criterion might include "high diversity" as a required attribute, in which case the scenario would satisfy this criterion.

To use Risk-MAP, one normally loads a specific knowledge base that has been previously created. For purposes of illustration, a demonstration knowledge base called BIO-Logic was developed for addressing the possible risks of introducing genetically-modified organisms into the environment. The knowledge was extracted from various scientific reports and symposia that have focused on this important subject over the past few years.

6.2. Knowledge Representation

The BIO-Logic knowledge base is organized as a hierarchy of classes, moving from the general to the specific, as shown in Figure 3. The user can easily browse through the current knowledge base by selecting the classes they wish to focus on. There are four major types of knowledge represented:

Concepts: The classes of concepts defined in BIO-Logic include organisms, ecosystems, modifications, proliferation stages, and outcomes. Each class is subdivided into more specific subclasses; for example higher organisms are divided into insects, plants, animals, and humans.

Attributes:

Attributes are properties that apply to concepts. For example, attributes that apply to bacteria might include host range, genetic stability, donor organism type, modification, pathogenicity, infectivity, etc. Attributes can be either qualitative or quantitative.

Scenarios:

Scenarios are descriptions of particular situations to be evaluated. For example, in BIO-Logic scenarios represent proposed applications of genetically modified organisms. A scenario might have the following concepts defined as components:

bacterium: rhizobium
ecosystem: soil ecosystem
beneficial outcome: plant growth enhancement
possible risk: displacement of indigenous species

Criteria:

Criteria are sets of conditions that are relevant to decision-making or priority-setting, including regulatory constraints, review policies, or risk indicators. Criteria are defined by stipulating the attribute conditions that must be present (or absent) in the scenario of concern. For example, a criterion might require that the microorganism be a nonpathogen.

The user can also specify logical inference rules that will deduce attribute values. For

Figure 3

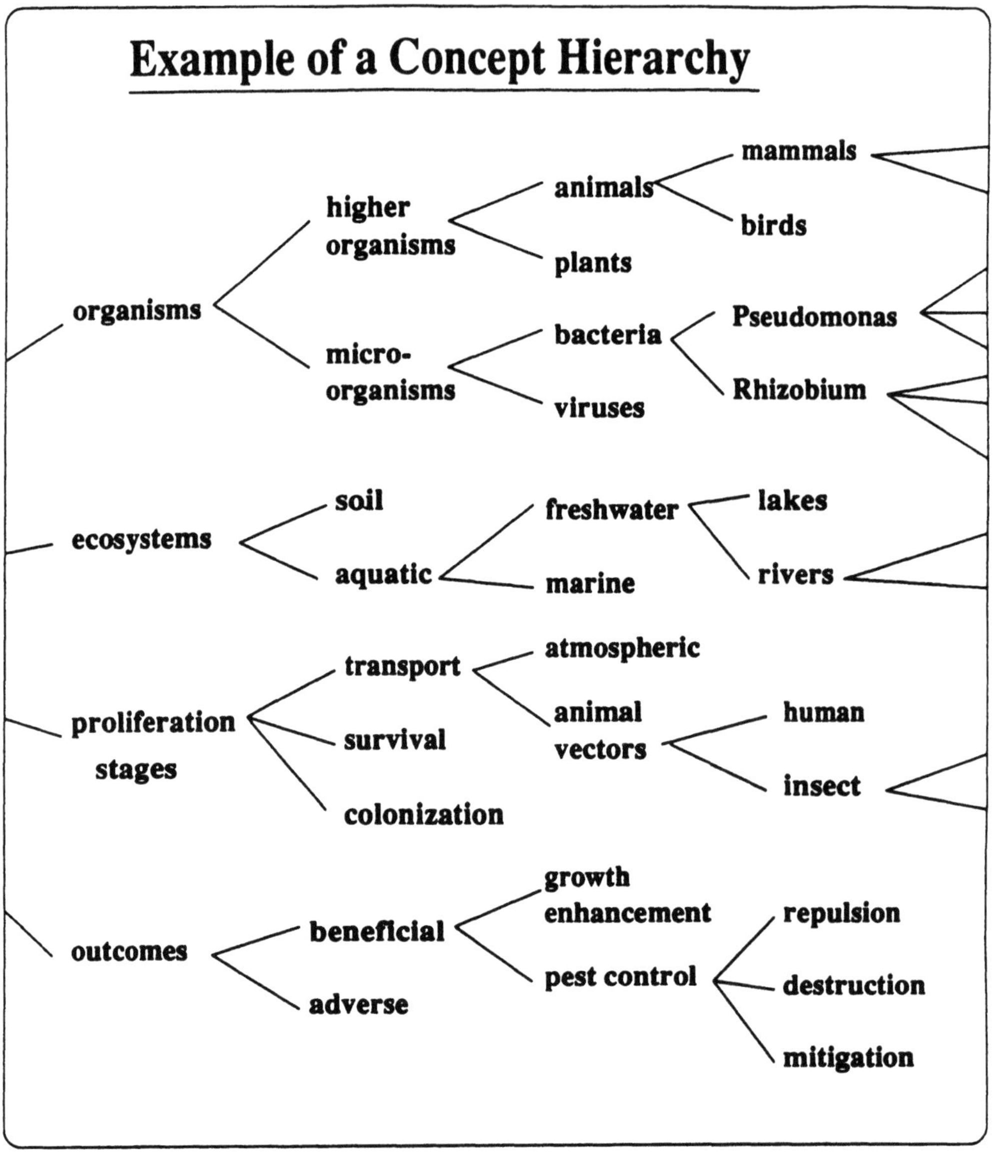
Example of a Concept Hierarchy
organisms
higher organisms
micro-organisms
animals
plants
bacteria
viruses
mammals
birds
Pseudomonas
Rhizobium
ecosystems
soil
aquatic
freshwater
marine
lakes
rivers
proliferation stages
transport
survival
colonization
atmospheric
animal vectors
human
insect
outcomes
beneficial
adverse
growth enhancement
pest control
repulsion
destruction
mitigation

example, a sample rule in BIO-Logic is:

If the donor organism of X is A
and A is known to be pathogenic
then X may exhibit pathogenic traits.

While browsing, the user can modify or delete anything in the knowledge base. Moreover, a unique aspect of Risk-MAP, in contrast to previous software tools, is that it allows the user to incrementally add new knowledge to the knowledge base. The user simply indicates what type of knowledge they wish to define, and Risk-MAP guides them in expressing it.

6.3. Screening and Evaluating Scenarios

Risk-MAP can screen any selected scenarios against any selected criteria. Risk-MAP tests each scenario to see if it matches the required or excluded attributes, provides ongoing explanation of the findings, assesses the degree of of certainty that the criteria are satisfied, and displays an appropriate message. If there are gaps in the knowledge base, then Risk-MAP will try relevant inference rules or query the user to determine values of unknown attributes. Moreover, for any set of criteria, Risk-MAP can rank several scenarios based upon criteria weights assigned by the user, and determine the highest-ranked scenario.

Finally, Risk-MAP can help in making qualitative comparisons among scenarios. For purposes of understanding the potential risks associated with a given scenario, it is often helpful to use comparisons with previously-encountered scenarios. For example, if the scenario being evaluated involves an application of modified "rhizobium" in a "soil ecosystem," Risk-MAP can seek all scenarios in BIO-Logic that also include either of the concepts "soil ecosystem" or "rhizobium".

Risk-MAP can then examine each of these comparable scenarios to see if its attributes are similar to the scenario of interest. For example, the "host range" attribute can be tested for each scenario that includes "soil ecosystem," to see whether the organism in question has a host range similar to the modified rhizobium. Scenarios that are discovered in this way by Risk-MAP can then be screened or ranked against the scenario of interest to assist in decisions about acceptability or appropriate safety assurance measures.

7. Conclusions

This paper has suggested that effective safety assurance for environmental introductions of genetically-engineered microorganisms requires a comprehensive scientific and regulatory knowledge base and a methodology for interpreting and applying that knowledge. Computer-based systems using artificial intelligence can provide reliable and consistent support to humans in the difficult task of evaluating proposed introductions, and particularly in the qualitative screening exercise that is typically the first stage of review.

For many innovative types of environmental introductions, empirical risk assessment methods will be impractical due to the lack of an adequate data base of organism characteristics and interactions with the target ecosystem. Similarly, model-based assessment will be difficult or impossible due to the current lack of adequate predictive models of environmental outcomes for specific ecosystems. Therefore, for the foreseeable future, biotechnology risk assessment will rely heavily upon qualitative screening methods such as those currently practiced by the National Institutes of Health Recombinant DNA Advisory Committee.

Knowledge systems can support this process by:

- permitting the capture, maintenance, and distribution of a large body of regarding technical, economic, or scientific knowledge

- assisting in screening proposed environmental introductions for characteristics that might conceivably lead to adverse outcomes

- helping to implement generic guidelines by systematically and exhaustively reviewing all proposed introductions

- recognizing the appropriate classification for a proposed introduction and triggering the required review procedures

- executing complex logical schemes that support humans throughout the various stages of risk assessment and risk management

In order to demonstrate the application of knowledge systems a simple protoype system was constructed that assists with qualitative and quantitative decision-making in the safety assurance process. The foundation of this system is a knowledge base called BIO-Logic that reflects the basic biological concepts underlying environmental introductions. This prototype illustrates the use of "intelligent decision support systems," which will become increasingly commonplace over the next few years.

Though risk assessment for environmental introductions of genetically engineered organisms is at an early stage of development, qualitative screening methods can provide adequate safety assurance without requiring elaborate mathematical models. Eventually, as more rigorous analysis methods are developed and as the available data base expands, formal risk assessments will be possible for a broad range of biotechnology applications. This paper has suggested that analytical logic schemes can provide decision-makers with guidelines for invoking the full range of quantitative and qualitative methods as necessary.

8. References

[1] Fiksel, J. and V.T. Covello, "The Suitability and Applicability of Risk Assessment Methods for Environmental Applications of Biotechnology," in J. Fiksel and V. Covello (eds.), "Biotechnology Risk Assessment," Pergamon Press, 1986, pp. 1-34.

[2] NATO Advanced Research Workshop Summary Report, "Recommendations for a Scientific Approach to Safety Assurance for Environmental Introductions of Genetically-Engineered Organisms," Rome, Italy, June 6-10, 1987. (not yet published)

[3] National Academy of Sciences, "Risk Assessment in the Federal Government: Managing the Process," Washington, DC, 1983.

[4] Merkhofer, M. and V. Covello, "Risk Assessment and Risk Assessment Methods:

The State of the Art," Plenum, N.Y., 1986.

[5] Russell, M. and M. Gruber, "Risk Assessment in Environmental Policy-Making," Science, Vol. 236, 17 April, 1987, pp. 286-290.

[6] King, D. and R. Morgan, "An Overview of Knowledge-Based Systems," Data Processing Management, Auerbach, 1986.

[7] Hayes-Roth, F., D.A. Waterman, and D.B. Lenat, Building Expert Systems, Addison-Wesley, Reading, MA (1983).

[8] Fiksel, J., "Artificial Intelligence: Software Reasons to Analyze Risks," Safety and Health, National Safety Council, March, 1987, pp. 64-66.

[9] Rossman, L.A., "The Use of Knowledge-Based Systems in RCRA and Superfund Programs," Paper presented at NCASI Meeting, October, 1986, U.S. E.P.A., Cincinnati, Ohio.

ASSESSING HUMAN HEALTH RISKS OF ENVIRONMENTALLY RELEASED, GENETICALLY ENGINEERED MICROORGANISMS

C.A. Franklin and N.J. Previsich
Environmental Health Directorate
Department of National Health and Welfare
Ottawa, Ontario
K1A 0L2, Canada

LEGISLATIVE FRAMEWORK

Many countries have identified biotechnology as a priority technology and are actively promoting its research and development. The prospect of rapid advances towards commercialization for a wide range of products has focused attention on the potential health hazards and environmental impacts. The adequacy of existing legislation to regulate biotechnology from the perspectives of mandate and appropriateness of existing data requirements is being assessed.

It was reported by OECD that many countries were planning to use existing environmental laws for regulating genetically engineered organisms and their products (1). Some countries have drafted guidelines that are intended to assist industry in obtaining the appropriate clearance to release genetically engineered organisms and to help regulators evaluate the associated risks (2-6). The biotechnology laws and regulations of Japan, France, Germany and European communities have been reviewed (7).

The advantages of using existing legislation include knowledge of the effectiveness of the existing legislation, its strengths and weaknesses and provision of an immediate capability to regulate. However, there are disadvantages in trying to utilize legislation that was developed primarily for chemicals. Living organisms pose unique problems and will require new tools to assess safety. There are inconsistencies in the existing data requirements for different types of chemicals. Products, such as drugs, direct food additives and pesticides require extensive safety and efficacy testing with premarket review prior to commercialization. Other chemicals are not subjected to such stringent premarket

NATO ASI Series, Vol. G18
Safety Assurance for Environmental Introductions of Genetically-Engineered Organisms
Edited by J. Fiksel and V.T. Covello

evaluation and may only require a minimum data package prior to clearance (premanufacture notices under Toxic Substances Control Act (TSCA) in the U.S. or notification after sales reach a specified volume under the Environmental Contaminants (EC) Act in Canada). It would be unfortunate if these inconsistencies were extended to products of biotechnology.

In 1980, a Task Force on Biotechnology was established to advise the federal government on effective strategies to promote biotechnology in Canada (8). Federal monies were committed to encourage industrial and provincial government investment and to expand federal research programs (9). There has been significant activity by Canadian industries in the development of biotechnology products. A survey conducted by the National Biotechnology Advisory Committee, which advises the Minister of State for Science and Technology, identified 110 organizations engaged in commercial operations involving biotechnology research, development or manufacturing in 1985 (10). Human health care products, food and beverage industries and agriculture were reported as major areas of activity. Seventy per cent of total R & D expenditures in biotechnology were in the health care and agricultural sectors.

The primary federal acts in Canada that can be used for the regulation of environmentally released products of biotechnology are the Food and Drugs Act (F & D Act), the Pest Control Products Act (PCP Act), and new legislation, the Canadian Environmental Protection Act (CEPA, Bill C-74), that replaces the Environmental Contaminants Act (EC Act). Products of biotechnology, including chemicals produced by organisms, as well as the organisms themselves will be regulated under CEPA. This Act will regulate products not covered by other legislation.

RISK ASSESSMENT

One basic question is whether existing models developed for the risk assessment of chemicals are suitable for assessing the risks of genetically engineered organisms. The differences in approaches in the risk assessment of chemicals (11) are most apparent in the delineation between risk assessment and risk management (Fig. 1). In all of the models presented, with the exception of NRC/EPA (1983), risk assessment has 3

SCOPE (1980)	NRC/EPA (1983)	ROYAL SOCIETY (1983)	ICTC (1984)	WHO (1985)
RISK IDENTIFICATION	RESEARCH HAZARD IDENTIFICATION	RISK ESTIMATION	HAZARD IDENTIFICATION	HAZARD IDENTIFICATION
RISK ESTIMATION	DOSE-RESPONSE ASSESSMENT EXPOSURE ASSESSMENT RISK CHARACTERIZATION		RISK ESTIMATION	RISK ESTIMATION
RISK EVALUATION	DEVELOPMENT OF REGULATORY OPTIONS EVALUATION OF OPTIONS	RISK EVALUATION	DEVELOPMENT OF ALTERNATIVE COURSES OF ACTION	RISK EVALUATION
RISK MANAGEMENT	DECISIONS AND ACTIONS	RISK MANAGEMENT	IMPLEMENTATION MONITORING AND EVALUATION REVIEW	RISK MANAGEMENT

Figure 1. Comparison of models for risk assessment (above dotted line) and risk management (below dotted line). Adapted from (11).

basic components. Risk or hazard identification is the initial step and relies heavily upon toxicology and epidemiology to provide the data on health outcomes. The next step is risk estimation in which the probability of occurrence of an event or activity is assessed by statistically analyzing the toxicology and epidemiology data and the level of human exposure. The third step is risk evaluation or option analysis and may involve consideration of program objectives, current institutional policies and the regulatory environment. The NRC/EPA model considers this option analysis section to be part of risk management. The other models consider risk management to be the selection and implementation of a strategy for control of a risk, followed by monitoring and evaluation of the effectiveness of the strategy.

A brief overview of the approaches used for assessing human health risk to chemicals and to naturally occurring microbes will assist in providing a perspective on risk assessment for genetically engineered organism.

Details of steps that are used in assessing the risk of chemicals to human health are shown in Fig. 2. The first two, hazard identification and risk estimation require the development of databases on toxicity and exposure. Historically the hazard of a chemical has been assessed through the use of predictive toxicity testing in a variety of animal species. The types of studies that are conducted include acute (single exposure), subchronic (multiple exposure) and chronic (lifetime exposure) studies as well as studies to assess reproductive effects, teratogenic effects, and mutagenic/carcinogenic effects (Table 1). The intent of these studies is

Table 1. Toxicity studies used to assess human health hazards of chemicals

ACUTE	SUBCHRONIC	LONG TERM AND SPECIAL STUDIES
Acute - Oral - Dermal - Inhalation	Oral - 90 day - 12 month (dog)	Chronic Feeding Oncogenicity
Irritation - Dermal - Eye (when necessary)	Dermal - 90 day Inhalation - 90 day	Pharmacokinetic (absorption distribution excretion and metabolism) Mutagenicity
Sensitization	Delayed Neurotoxicity	Teratology
Delayed Neurotoxicity		Multigeneration Reproduction

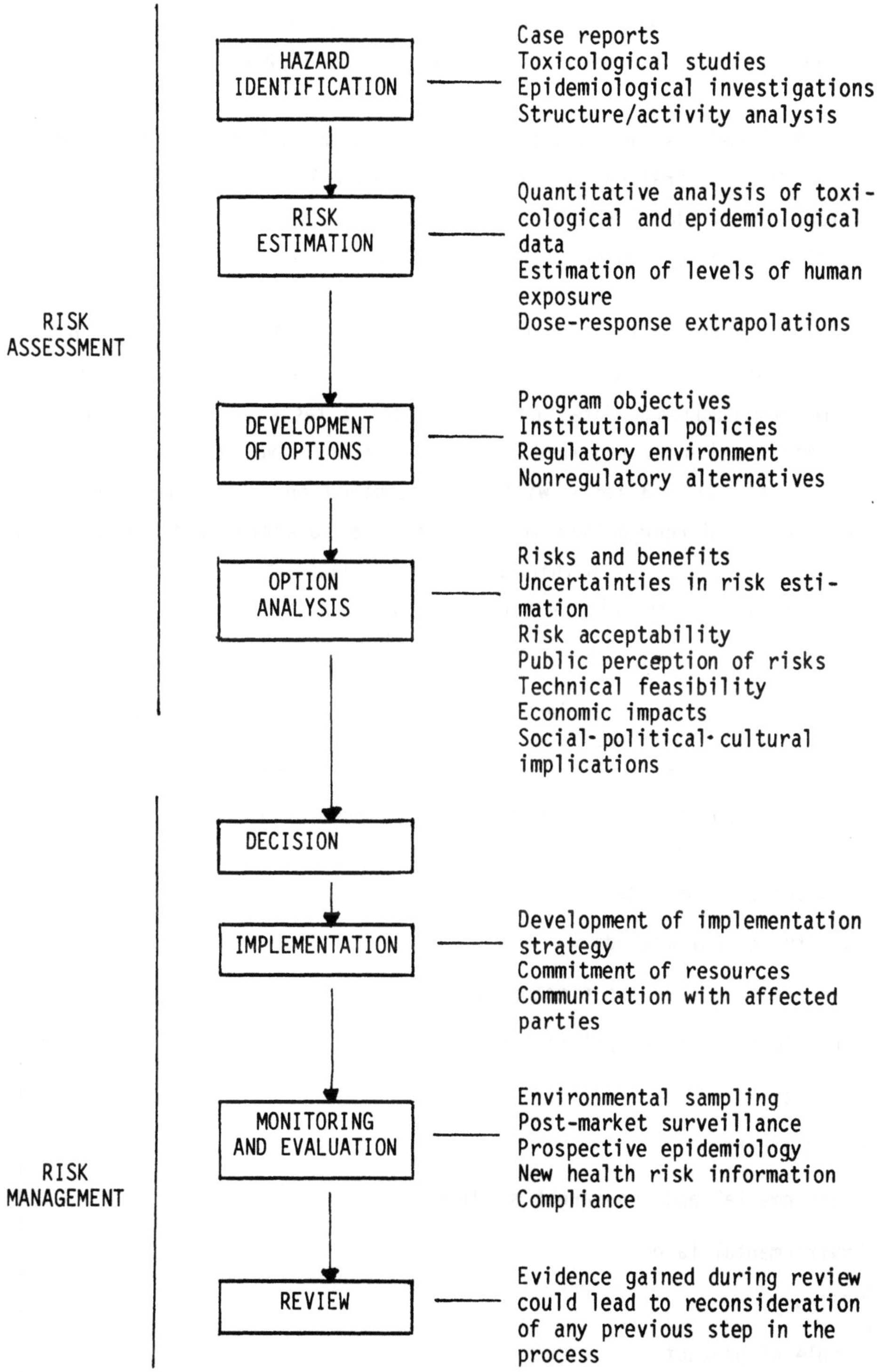

Figure 2. Details of a risk assessment/risk management model suitable for assessment of chemicals.

to identify all possible toxicological end points and to establish the dosage at which no adverse effects are observed (NOAEL).

A tiered system is used for assessing the safety of microorganisms prior to registration as pesticides. Guidelines have been developed which acknowledge that microorganisms are inherently different from chemicals and so require different safety testing (Table 2).

A general outline of the tier system currently being considered for use in Canada is shown in Table 3. All tests are to be conducted using a maximum challenge dose and if all Tier 1 tests are negative then no further testing is required. If any are positive, then the necessary Tier 2 and Tier 3 tests will be determined on a case-by-case basis. Testing of microorganisms must assess the potential for organisms to infect and/or persist within tissues, to elicit toxic reactions and to cause ocular or dermal irritation or hypersensitivity.

Table 2. General categories of data required for registration of microbial pesticides

Specifications (strain, identification, amount of active ingredient)
Manufacturing methods
Quality control methods
Impurities or extraneous materials
Certification of ingredient limit
Toxicity
Residue data
Environmental and non-target studies
Environmental fate
Efficacy
Sample of product

Table 3. Studies used to assess hazards of microbial pesticides

TIER 1	TIER 2*	TIER 3**
Acute Infectivity/Toxicity - Oral - Dermal - Inhalation	Quantification of Persistence (most appropriate route) Short term toxicity (90 day most appropriate route)	Chronic oral Oncogenicity Teratogenicity Reproduction
Acute Infectivity (normal and immuno-suppressed animals) - Intravenous - Intraperitoneal - Intracerebral	Oncogenicity Reproduction Teratogenicity Mutagenicity	
Irritation - Dermal - Eye (when necessary)		
Delayed Hypersensitivity		
Tissue Culture (viruses) Genotoxicity testing for toxins and by-products		

* neurotropic agents
** determined case-by-case

Combined acute infectivity/toxicity tests using oral, dermal and inhalation routes of exposure must be conducted on all organisms. The animals are observed clinically and examined daily for 28 days after dosing. The standard parameters for toxicity (biochemistry, hematology, histopathology) are measured. In addition, tissue cultures of organs should be taken to isolate viable organisms. Acute infectivity is further assessed by dosing intravenously (bacteria and viruses) and intraperitoneally (fungi and protozoan) using normal and immuno-suppressed animals. If a toxin is produced it must be run through a genotoxic screen and a long term oncogenicity study and possibly other special studies may be required. A sensitive method for detecting infectivity and DNA interaction in viruses is tissue culture testing (in vitro) using various cell lines. Delayed hypersensitivity is another endpoint that must be assessed since microbes and their products are potential allergens. It is advised that workers undergo a yearly medical examination. Any relevant and significant findings related to exposure should be reported.

Since the naturally occurring microorganisms currently registered for environmental applications have been assessed as being non-infective, there is no need to assess exposure or to proceed into Tier 2 testing. This abbreviated approach to testing has also been used with chemicals that are deemed to be non-toxic.

The question that most regulatory agencies have had to address is whether genetically engineered organisms are potentially more hazardous than naturally occurring organisms. One approach to regulating genetically engineered organisms is to consider that the new organism does not differ from its naturally occurring parents unless specific parameters or triggers are exceeded. In the United States, it has been proposed that genetically engineered micro-organisms that contain material from dissimilar source organisms (intergeneric organism), that are known pathogens or contain material from pathogens, or are non-indigenous should be subjected to some level of testing. The U.S. Environmental Protection Agency (EPA) has developed a scheme for regulating both naturally occurring organisms and genetically engineered organisms based on these parameters and the data that may be required are divided into Level I and Level II categories (Table 4). A trigger level has also been set for applications that will be less than or greater than 10 acres. This approach differs from both the one used to assess chemicals where all endpoints must be tested for each chemical and for microbial products where a tier testing approach is utilized.

Theoretically the use of triggers to identify "new" organisms that would require testing has merit. However there are serious limitations in our ability to determine whether the new organism remains true to its origins or whether it takes on its own unique characteristics. In addition to this, there are some fundamental problems with applying each of the triggers for data exemption.

1. Intergeneric Transfer Trigger. It is prudent to have some level of testing on the new organism since the probability of intergeneric transfer occurring naturally is unknown. Even simple transfers and additions or deletions should be looked at in the initial phases to build a data base for future possible exemptions.

Table 4. General data requirements (U.S. EPA) for genetically engineered organisms to be applied in the environment

Level I Information (abbreviated review)
. identity of microorganism (characteristics and means and limits of detection)
. description of the natural habitat of microorganism of parental stains (predators, parasites and competitors)
. host range of parental strain or non-indigenous microorganism
. environmental competitiveness
. methods used to engineer: identity and location of rearranged or inserted/deleted gene, description of new trait, information on potential for genetic transfer and exchange, genetic stability of inserted sequence
. description of proposed testing program
Level II Information
. background information on microorganism
. identity
. description of habitat
. host range, especially infectivity and pathogenicity to non-target organisms
. survival and ability to increase
. detailed information on identity, genetic transfer, application size
. monitoring
. disposal
. evaluation of potential health effects (Guidelines for Biorationals)

2. Pathogen Trigger. A pathogen has been defined as a virus or organism that has the ability to cause disease in other living organisms and an organism will be identified as a pathogen if: (1) it belongs to pathogenic species or to a species containing pathogenic strains. Exceptions are organisms generally recognized as non-pathogenic i.e. Escherichia coli K-12, Bacillus subtilis, Lactobacillus acidophilus and Saccharomyces sp.; (2) it has been derived from a pathogen or has been deliberately engineered from a pathogen (exceptions are transfer of well-characterized non-coding regions); (3) no data exist to the contrary, (4) it contains genetic material from a virus (15). Unfortunately the current state of our knowledge regarding taxonomy is unclear and it may not be easy to identify whether the organism belongs to a pathogenic species (16). It is difficult to be able to properly assess whether the new organism is considered indigenous or non-indigenous.

3. Indigenous Trigger. Equally difficult to assess without testing are the impacts of increasing the numbers of an organism well beyond its naturally occurring density and the effect of dissemination by unnatural routes. It may be that as we accumulate a body of knowledge on micro-organisms we will be better able to predict the outcome of changes to existing organisms.

EXPOSURE ASSESSMENT

Although exposure assessment is one of the major components of risk assessment, it has not been as well characterized as the hazard component. There is even less knowledge of the exposure to workers or the general population resulting from the use of microbial pesticides since most of these products were presumed to be safe. The primary purpose for quantifying exposure is to enable comparison of the exposure level resulting from use of the product to the level known to cause toxic effects in the test animals. If there is an adequate margin between these levels, it may be concluded that the product can be used safely.

Exposure to humans can be estimated through environmental and biological monitoring (17, 18). The general environment (air, water, food) can be monitored and the concentration of the chemical in the ambient sample then extrapolated to estimate exposure to the individual (Fig. 3). This type of sampling provides the least accurate estimate since it presumes uniform distribution in the environment and exposure. A second type of sampling that has been used involves the placement of dosimeters on the individual enabling estimation of the actual amount of material that contacts the skin, or is potentially inhaled. This provides a more accurate estimation than ambient sampling but does not give an estimate of the integrated dose that is absorbed through the skin, lungs or gastrointestinal tract. Monitoring of a biological response or excretion of metabolites provides the most accurate estimation of absorbed dose and relates best to the toxic dose.

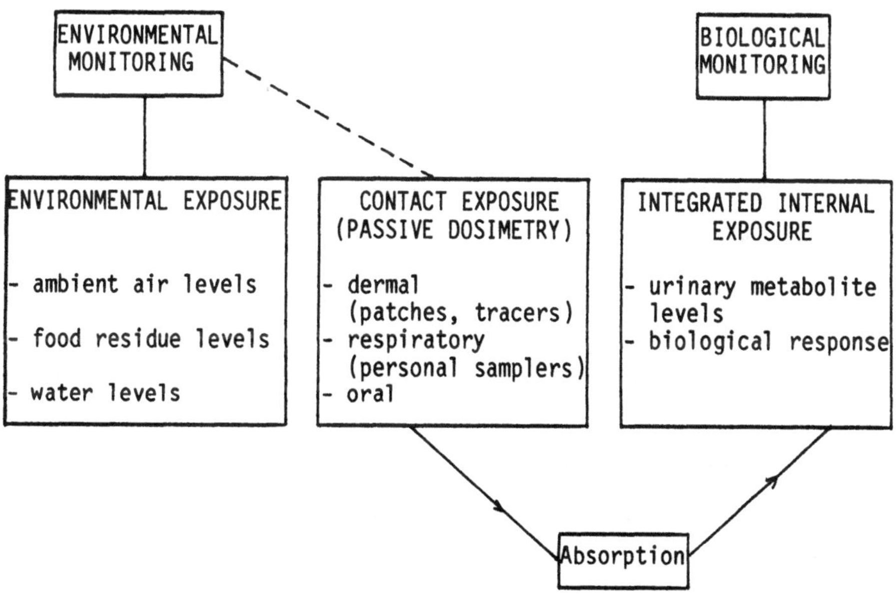

Figure 3. Components of exposure to workers and bystanders

FACTORS AFFECTING EXPOSURE

When trying to estimate exposure to the general population following use of a microbial product there are a number of additional factors that must be considered such as dispersion, number of organisms that are emitted and infectivity.

1. Dispersion. It has been estimated that microbes can be dispersed in air for distances from hundreds to thousands of kilometers. At a windspeed of 5 m/s a 1 μ and a 5 μ bacterium would be transported 170 and 7 km while settling 1 m (19). Bacteria associated with aerosols or particles would settle faster.

Studies conducted in Quebec following application of *Bacillus thuringiensis kurstaki* (B.t.) at the rate of 1.3 x 10^8 to 2.6 x 10^8 CFU per 0.093m^2 showed measurable levels of B.t. in air for up to 50 days (20). The values were extremely variable and the maximum levels measured at 7 sites ranged from 7.6 to 938 CFU/m^3. It was felt that the upper limit was spurious due to contamination and that the next highest level of 29 CFU/m^3 was more reasonable. There was only a weak correlation observed with distance from application site (Table 5). Levels in water after spraying are shown in Table 6.

Table 5. Maximum concentration of *Bacillus thuringiensis k* in 288 air samples from 7 sites following aerial application in Quebec

Site	Day (Max)	CFU/L/min	CFU/m^3
1	10	10.5	7.6
2	10	42.2	29.4
3	11	1365.1	938.2 *
4	34	87.5	11.4
5	37	21.0	14.5
6	7	14.4	10.0
7	36	17.36	12.1

* contaminated

Comment

- application rate was 1.3 x 10^8 to 2.6 x 10^8 CFU/0.93 m^2
- 84% were < 1 CFU/L/min
- relatively weak correlation with distance from application site

Table 6. Aerial spraying of Bacillus thuringiensis k in Quebec

TIME AFTER SPRAYING	B.t. CONCENTRATION (CFU x 10^4/L) IN STREAMS	
20-120 minutes	58-610	(90 samples)
11 to 50 days	0.22-4.6	(90 samples)

2. Number of organisms emitted. The number of organisms that may be emitted during various activities (21) are summarized in Table 7. It has been estimated that a BL1 containment laboratory could release up to 4 x 10^9 microbes. Since BL1 has been designated as the lowest level of containment permitted for working with recombinant DNA it is not considered to be an environmental release. It was proposed that concentrations of organisms higher than this be considered environmental releases and thus subject to review. As shown in Table 7, this would include activities such as conventional types of greenhouses, agricultural field tests and waste clean up.

Table 7. Potential emissions of microorganisms

USE	NO. OF MICROBES IN INOCULUM	NO. OF MICROBES ESCAPING	ENVIRONMENTAL RELEASE
BL1 Waste stream	n.a.	2×10^8 to 4×10^9	no
Greenhouse			
Conventional	10^{12}	10^{11}	yes
BL1-type	10^{12}	10^8 to 10^9	no
Agricultural Field Test (Ice minus, B.t. endotoxin)	10^{12}	10^{12}	yes
Waste cleanup	2 x 10^{14}	2×10^{14}	yes

adapted from (21)

3. Infectivity. These examples of dispersion and the calculation of the number of organisms that escape from the use site illustrate the point that exposure to humans will occur following use of genetically engineered organisms. Although the regulatory position at this time appears to preclude the use of pathogenic organisms for environmental applications, it would be extremely useful to know whether the maximum challenge dose used for testing is sufficiently larger than environmental levels that will occur. As shown in Table 8 there is a very wide dose range (22) over which known pathogens are infective and this type of information may be useful in estimating potential risks for genetically engineered organisms.

The route of exposure to the organism can also affect the outcome; bubonic plague occurs if the bacteria reach the lymph nodes and pneumonic plague if the bacteria are inhaled or reach the lungs (23).

Table 8. Doses of Selected Pathogens causing a 50% infection rate

ORGANISM	50% INFECTIVE DOSE	ADMINISTRATIVE
Salmonella spp	10^5-10^8	Oral
Shigella	10-100	Oral
Giardia lambia	1-10	Oral
Adenovirus 4	1 (Tissue culture infective dose)	Nasal

adapted from (22)

CONCLUSIONS

1. Safety Assurance vs Risk Assessment. It has been suggested that genetically engineered organisms that do not set off the triggers could be considered "safe" and not have to be tested. Although this might be a useful goal to strive for, this approach goes against the requirements of existing legislation in many countries where safety assurances must be given on a product by product basis. There are also sufficient

uncertainties that would preclude the use of this approach at this time. However, it may be a goal to work towards.

2. Case-by-case Approach. General guidelines can be developed for genetically engineered organisms as they have been for chemicals. Nevertheless, each product must be assessed against these guidelines and any additional testing that is warranted conducted. As our knowledge increases, similar types of organisms may be judged acceptable with less data or in fact may be considered exempt on the basis of relatedness to similar organisms considered acceptable ("safe"). It should be emphasized that this generic or structure - activity approach for evaluating chemicals was not as successful as originally anticipated, and it is likely that similar problems may exist when applying a generic approach to organisms.

3. Development of Databases. More emphasis should be placed on trying to obtain as much information as possible on the behaviour of non-engineered organisms released to the environment and using these data to answer the many conjectural type of questions on survival, proliferation, establishment and genetic transfer.

In summary it would be unfortunate if the regulatory climate became unduly stifling to innovation; however it could be far more serious if inappropriate organisms were environmentally released.

Acknowledgement

The authors wish to thank Brenda MacDonald for valuable input in the section on toxicity testing for microbial pesticides and for reviewing the manuscript.

REFERENCES

1. Organization for Economic Cooperation and Development. "Safety and Regulations in Biotechnology", Draft Report, DSTI/SPR/85.23, June 18, 1985.

2. Organization for Economic Cooperation and Development. "Recombinant DNA Safety Considerations", Paris, 1986.

3. "The Planned Release of Live Organisms Modified by Recombinant DNA Techniques: Points to Consider in the Preparation of Proposals for Genetic Assessment". Recombinant DNA Monitoring Committee, Australian Department of Industry, Technology and Commerce, May, 1985.

4. "Draft Recommendations for the Field Testing and Release of Genetically Modified Organisms in New Zealand.", Ministry of Science and Technology. September, 1986.

5. "Guidelines for the Application of Recombinant DNA Organisms in Agriculture, Forestry, Fisheries, Food Industry and Other Related Industries: A Proposal", Ministry of Agriculture, Forestry and Fisheries, Japan, December 18, 1986.

6. "U.K. Issues Field Testing Guidelines" Recombinant DNA Tech. Bull. 9 (2), 1986.

7. Review and Analysis of International Biotechnology Regulations. Prepared by Arthur D. Little Inc. ADL 54132, May 1986.

8. "Task Force on Biotechnology: A Development Plan for Canada", Canadian Department of Supply and Services, 1981, ST 31-9/1981-E.

9. Franklin, C.A., Nestmann, E.R. and L. Ritter. "Risk Assessment and Regulation of Genetically Engineered Products", Biotechnology in Agricultural Chemistry, American Chemical Society, 1987.

10. "1986 Canadian Biotechnology Sourcebook", Prepared for the National Biotechnology Advisory Committee, Ministry of State for Science and Technology, August, 1986.

11 Krewski, D. and Birkwood, P.L. Risk Assessment and Risk Management. Risk Abstr. 4(2): 53-61, 1987.

12. Guidelines for Registering Pesticides in the United States. Subpart M Data Requirements for Biorational Pesticides. Environmental Protection Agency, 1980.

13. Bulletin of the World Health Organization 1981, 59/6, 857.

14. Pesticide Assessment Guidelines. Subdivision M: Biorational Pesticides, Environmental Protection Agency, 1982.

15. "Coordinated Framework for Regulation of Biotechnology" Office of Science and Technology, Part II, Federal Register, June 26, 1986.

16. Strauss, H., Hattis, D., Page, G., Harrison, K., Vogel, S., and Caldart, C. "Genetically - Engineered Microorganisms: I. Identification, Classification, and Strain History", Recombinant DNA Tech. Bull. 9 (1), 1986.

17. Franklin, C.A., Muir N.I. and Greenhalgh R. In Pesticide Residues and Exposure, ACS Symposium Series 182, 157-168, 1982.

18. Franklin, C.A., Muir, N.I. and R.P. Moody. The Use of Biological Monitoring in the Estimation of Exposure During the Application of Pesticides, Toxicol. Letters. 33: 127-136, 1986.

19. Barnthouse, L.W. and A.V. Palumbo, "Assessing the Transport and Fate of Bioengineered Microorganisms in the Environment", In: The Suitability and Applicability of Risk Assessment Methods for Environmental Applications of Biotechnology. (ed) Covello, V.T. and J.R. Fiksel, National Science Foundation, Washington, D.C., August, 1985.

20. Major, L., Rousseau, G. and Cardinal, P. "Environmental Monitoring of the Aerial Spraying of Insecticides Against the Spruce Budworm in Quebec, 1985: Concentration of Biological Insecticides in the Air of Seven Municipalities in the Vicinity of Treatment Areas". Ministère de l'Energie et des Ressources, Québec. No. 624, Apr. 16, 1986.

21. Strauss, H.G. "How Many Microbes Really Constitute Environmental Release", Bio/technology, Vol. 5, NO. 3, March 1987.

22. Sorber, C.A., "Public Health Aspects of Agricultural Reuse Applications of Wastewater", Am. Soc. Chem. Eng. News. 57: 4-13, 1982.

23. Lincoln, D. Risk Assessment of Microbial Application in Biotechnology in Hazard Assessment of Chemicals Vol. 5 199-232. ed. J. Soxena Hemisphere Pub. Corp., 1987.

RISK ASSESSMENT STRATEGIES FOR BIOTECHNOLOGY

Curtis C. Travis and Holly A. Hattemer-Frey
Office of Risk Analysis
Health and Safety Research Division
Oak Ridge National Laboratory*
P. O. Box 2008, Building 4500S
Oak Ridge, Tennessee 37831-6109

*Research was sponsored by the U. S. Environmental Protection Agency under Martin Marietta Energy Systems, Inc., Contract No. DE-AC05-84OR21400 with the U. S. Department of Energy. Accordingly, the U. S. Government retains a nonexclusive, royalty-free license to publish or reproduce the published form of this manuscript, or allow others to do so, for U. S. Government purposes.

1. INTRODUCTION

Genetic engineering offers the promise of major advances in many areas of science, including agriculture, medicine, and chemical manufacturing [1]. Products of this technology, however, may also pose unknown risks to human health and the environment. Recent surveys show that while 62% of the public believe that genetic engineering will improve the quality of life, 52% think that it is inherently dangerous [2]. As with any new product, anticipated benefits must be balanced with potential risks. Before this task can be accomplished, however, a framework must be established to assess the risks associated with deliberate environmental releases of genetically altered organisms.

NATO ASI Series, Vol. G18
Safety Assurance for Environmental Introductions of Genetically-Engineered Organisms
Edited by J. Fiksel and V. T. Covello

Sales of biotechnology products in the U. S. are expected to reach several billion dollars by 1990 and as much as $40 billion by 2000 [3]. Despite this substantial projected growth, research on potential environmental and human health risks associated with deliberate releases of biotechnology products lags behind. A well-defined methodology for assessing these risks does not yet exist. This paper discusses the key risk-related issues surrounding bioengineered organisms and evaluates the appropriateness of applying current scientific risk assessment methodologies to biotechnology.

2. KEY ISSUES AFFECTING BIOENGINEERED MICROORGANISMS

Several concerns have been raised regarding the introduction of bioengineered organisms into the environment and the possibility that they will have deleterious effects on human health and the environment. Although these concerns heighten the controversy surrounding the development and use of biotechnology products, they must be addressed so that potential risks can be minimized and potential benefits maximized. The following questions are illustrative of the critical issues surrounding biotechnology.

2.1 Are Bioengineered Organisms Inherently Dangerous?

Biotechnology and genetic engineering are not new. The breeding and selection of plants and animals over the last 50 years to create specific, new traits is a well-established and accepted form of genetic engineering [4]. Furthermore, genetic manipulation has been practiced worldwide for more than a decade [5], during which time thousands of organisms have been artificially modified, and no tangible hazards have surfaced [6]. This historical experience led to the scientific consensus that genetically altered organisms are not inherently different from organisms produced by conventional methods and that they, consequently, do not pose unique environmental or human health hazards [6].

Genetic engineering procedures allow very precise alterations to be made, which are used to create an organism that will perform a specialized task after its release. These engineered specialists often have a reduced fitness (i. e., are at a competitive disadvantage), because most of their energy is devoted to performing the intended specialized task in lieu of basic survival tasks [7, 8]. As a result, they survive poorly when in competition with larger, established populations of naturally occurring organisms [7, 9, 10].

2.2 Can Biotechnology Alter an Organism in Wholly Unpredictable Ways?

The expression of a particular trait is controlled by multiple genes. Historical experience with plant and animal breeding experiments indicates that minor alterations in an organism's genetic makeup yield small changes in its characteristic behavior. Since genetic engineering usually involves the manipulation of a single gene, it is not likely to alter an organism's behavior in wholly unpredictable ways [11]. Hence, there is little evidence to suggest that bioengineered organisms will behave very differently from naturally occurring ones [6, 8].

2.3 Can Biotechnology Create Pathogens from Nonpathogens?

Studies with pathogens have shown that many varied and interacting traits are required for an organism to be pathogenic [8, 11]. These traits are typically controlled by multiple genes which do not function independently but rather make up an interactive, interdependent system [6]. A single gene transfer in a nonpathogenic organism is not likely to significantly enhance its pathogenic potential, because other traits necessary for pathogenic behavior would be lacking [4]. Hence, it is unlikely that the manipulation of a single gene will inadvertently turn a nonpathogenic organism into a pathogenic one [4, 6, 8].

2.4 Can Introduced Genes Spread to Other Organisms?

The exchange of genetic information in natural populations is ubiquitous and common [6]. Thus, it is reasonable to infer that innumerable genes have been transferred between a myriad of unrelated species over the last several hundred years, and yet no peculiar hazards or major ecological disruptions have occurred [4]. Despite the fact that large amounts of genetic material can be spread among unrelated populations in the presence of optimal selection pressures, scientific evidence indicates that such transfers usually yield unstable populations [6]. Again, genetically altered organisms rarely thrive in competition with larger, established populations of naturally occurring organisms [7, 9, 10].

3. A RISK ASSESSMENT METHODOLOGY FOR BIOTECHNOLOGY

Before any new chemical or organism, regardless of how it was developed, is released into the environment, expected environmental and economic benefits must be balanced with potential risks [11]. Assessing the risks of bioengineered organisms should be based on the nature of the organism and the environment into which it will be introduced and not on the process by which it was developed [8]. New products manufactured using new techniques do not necessarily require distinct risk assessment paradigms [4]. Because the risks associated with introducing bioengineered organisms into the environment are similar in principle to risks associated with releasing new, unmodified organisms or chemicals [6], current risk assessment methodologies for toxic chemicals and unaltered organisms can provide a framework for the assessment of biotechnology products [8,12]. These methodologies are based on the following components: (1) hazard assessment; (2) exposure assessment; (3) effects assessment; (4) risk management; and (5) risk communication.

Hazard assessment involves identifying potential adverse environmental and human health impacts resulting from intentional releases of genetically engineered organisms. Exposure assessment evaluates the risks associated with exposure to biotechnology products. Effects assessment characterizes the potential health effects and other consequences resulting from environmental releases of bioengineered organisms. Risk management seeks to determine what risks are reasonable or acceptable by comparing costs and benefits. Risk communication endeavors to inform and educate the public about potential hazards.

To evaluate potential risks associated with exposure to bioengineered microorganisms, it is essential to ascertain if the organism's intrinsic survival or pathogenic potential has been altered [13]. Knowledge of the original or parental organism is useful for assessing potential risks from exposure to engineered organisms, since genetically altered organisms resemble parental organisms in their ability to survive and persist in the environment [6, 13].

In addition to comparison with naturally occurring organisms, the use of mathematical models and simulations, model ecosystem experiments, and field tests enhance the risk assessment process [1,14]. Although many of the mathematical models available for analyzing the transport and fate of chemicals in the environment can be successfully applied to biotechnology products [1,15], risk assessment strategies for bioengineered organisms must ultimately be based on well-controlled, empirical studies. Aquatic and terrestrial model ecosystems (microcosms, mesocosms, growth chambers, and greenhouses) simulate complex environmental processes, so that population dynamics and ecosystem interactions can be studied. Model ecosystem experiments enable scientists and regulators: (1) to estimate the survivability of parental organisms in simulated, nonnatural environments [14]; (2) to evaluate potential adverse effects before organisms are introduced into the environment on a larger scale; and (3) to more reasonably predict whether significant risks will result from actual field releases of genetically altered organisms. More elaborate small-scale field tests can then be done to confirm model predictions. Better predictive capability with respect to a bioengineered organism's ultimate behavior and fate in the environment can only come from experimental research (including field tests) conducted on a site-specific and organism-by-organism basis [3, 7, 16]. This strategy provides a qualitative and quantitative basis for assessing the potential risks associated with deliberate environmental releases of bioengineered organisms.

4. FATE OF MICROORGANISMS IN THE ENVIRONMENT

Potential impacts of bioengineered microorganisms depend upon (1) their ability to grow, survive, and disperse in the environment, and (2) their direct and indirect effects on biota, ecosystems, and humans [17]. Although little is known about the survival and fate of bioengineered microorganisms in heterogeneous environments, a substantial body of knowledge characterizing the growth, survival, and dispersal of naturally occurring organisms and toxic chemicals already exists. These data can be used to predict the environmental behavior and fate of bioengineered microorganisms [6].

4.1 Survival and Growth

The survival of naturally occurring microorganisms depends upon: (1) temperature; (2) pH; (3) soil/sediment type and moisture content; (4) resistance to desiccation; and (5) adsorption to the soil surface [18]. Although bacterial survival generally declines during warmer months, adsorption to the soil surface can offer protection against desiccation and enhance survivability. The ability to tolerate differences in salinity and acidity (pH), soil type, and water availability varies widely among different species.

To survive, an organism must be able to coexist with competitors and predators and maintain itself during periods of low nutrient availability [18]. Factors favoring growth include: (1) the ability to withstand competition and predation pressures; (2) nutrient availability; and (3) efficiency of utilizing available nutrients [17]. Bacteria tend to exhibit one of three basic behavior patterns [19]. Some species decline rapidly and fail to persist. These species show a marked sensitivity to abiotic limiting factors (e. g., temperature, pH, etc.). Other populations decline at a steady but slow rate, and a small population ultimately survives. These species can compete for nutrients but are susceptible to other biotic limiting factors (e. g., predation and parasitism). Thirdly, some bacteria decline slowly initially but eventually reach a stable population size and persist for long periods of time. These species can tolerate both abiotic and biotic limiting factors. Liang *et al.* [19] concluded that, based on the behavior of naturally occurring microorganisms, most bioengineered organisms will not be able to tolerate the inherent biotic and abiotic stresses of their new environment.

4.2 Dispersal

Dispersal enables organisms to move from a poor or deteriorating environment to a more suitable one. All microorganisms seem capable of at least limited dispersal, although the modes and effectiveness of dispersal vary widely. Effective dispersal depends on: (1) source-pool size; (2) dispersal rate; and (3) organism survival during transport [15]. In general, the larger the source-pool size, the greater the number of organisms surviving transport, and the more effective the transmission. Short-range dispersal can occur via air, water, animal vectors, and direct contact with hosts [20]. Bacteria can disperse singly or attach to another particle or organism [20]. Solid particles, such as skin flakes or soil molecules, or aerosols serve as suitable dispersal agents. Particle size can also affect organism survival during dispersal. In general, organisms that attach to larger particles are more likely to survive, because larger particles contain more water [20]. Conversely, smaller particles settle out of the atmosphere more slowly and can be carried longer distances [15].

Water is also an effective short-range dispersal medium for several species of bacteria [20]. Although water dispersal is influenced primarily by the movement of water, bacteria can attach to particles or living organisms (such as algae or zooplankton) to enhance their dispersal potential. Thirdly, bacteria can be transported by animal vectors. For example, the *Plasmodium* microorganism, which causes malaria in humans, is transmitted by a mosquito, while the bacteria that causes squash wilt disease, *E. tracheiphila,* is vectored by a beetle.

Finally, some microorganisms are transmitted by direct contact with an infected host. This dispersal method is highly inefficient, however, and generally associated with the spread of infectious diseases [20].

Long-range dispersal, or passive airborne transmission of viable individuals, depends largely on weather conditions and the organism's ability to survive in the aerosolized state. Airborne microorganisms can be carried long distances by winds. Because long-range transport requires a large source pool, small desiccation-resistant organisms, and favorable meteorological conditions, few examples of long-range transport are reported in the literature [21].

5. APPLIED RISK ASSESSMENT AND BIOTECHNOLOGY: THE ICE MINUS RELEASE

One of the first formal risk assessments for a deliberate environmental release of a bioengineered organism was prepared by The University of California [22] for the proposed field test of a genetically altered organism. Frost damage worldwide is caused by a ubiquitous, naturally occurring organism, *Pseudonomas syringae.* Non-ice forming strains of *P. syringae* and ice nucleating strains that have mutated into non-iceforming strains also occur naturally but not in sufficient numbers to counteract the frost damage caused by ice nucleating strains. Researchers at the University of California used genetic engineering to transform an ice nucleating strain of *P. syringae* into one that prevents or reduces frost damage to crops. The genetically altered organism, called ice minus, differs from the parental strain in that it lacks a single gene that fosters ice formation on plant leaves but is similar to naturally occurring ice nucleating strains that have mutated into non-ice forming forms. Agricultural applications of ice minus could lengthen the growing season by increasing the productivity of frost-sensitive crops. Successful frost protection, however, depends on limiting the growth of the damaging ice nucleating organisms on plants [22]. Ice minus was precisely engineered to share the same ecological niche as naturally occurring ice nucleating strains. Although naturally occurring non-ice forming microbes offer some degree of frost protection, ice minus is best fitted to retard the growth of the naturally occurring ice nucleating microbes and minimize frost damage.

Proposed field tests of ice minus were designed to examine its fate and survivability in a competitive environment as well as its ability to prevent or reduce frost damage to crops. To assess the potential environmental and human health risks associated with introducing ice minus into the environment, the University of California [22] adopted a risk assessment strategy that integrated hazard, exposure, and effects assessments. The hazard assessment examined potential pathogenic and toxic effects of ice minus, while the exposure and effects assessments studied the environmental fate and impact of ice minus, including its survival in soil, its ability to compete and survive on indigenous flora, its colonization potential on non-target plants, and its potential to affect precipitation patterns. The risk assessment was based largely on laboratory studies involving both the naturally occurring and genetically altered strains of *P. syringae.* We briefly review the risk assessment that was completed before the proposed field test of ice minus took place.

5.1 Hazard Assessment

Although some strains of *P. syringae* are pathogenic to some plants species, the strain from which ice minus is derived has not been reported to be pathogenic to plants [22]. Furthermore, ice minus was tested for pathogenicity on more than 70 plant species, and results showed that it is not pathogenic to plants [22]. No strains of *P. syringae* are known to be pathogenic to animals or infectious to humans [22]. Despite the fact that humans and animals are exposed daily to *P. syringae* through contact with plants, its has never been associated with disease in animals or humans, because it cannot multiply at body temperature. Thus, it was concluded that ice minus would not be pathogenic to plants, animals, or humans, and the proposed field test held no foreseeable risks to human health or the environment [22,23].

5.2 Exposure Assessment

Before field applications of ice minus took place, extensive growth chamber and greenhouse experiments were done to verify that the survival and fate of ice minus in the environment could be predicted from knowledge of the behavior and fate of its parental strain [7]. These studies confirmed that during a small-scale application, ice minus would lack the numerical superiority to establish epiphytic dominance, because non-ice forming strains do not occur in sufficient numbers to exclude larger, established populations of ice nucleating organisms. Although ice minus was designed to limit the growth of its parental, ice nucleating strain, competition from other pre-existing microbial populations was expected to prevent the colonization of ice minus on neighboring plants.

The University of California [22] evaluated the growth and replication potential of ice minus and its parental strain in soil using assays that represented soil and plant debris expected to be present under field conditions. The concentration of both strains in soil declined logarithmically with time, and no viable organisms were found to persist in soil after 17 days regardless of initial population size [22]. These studies showed that ice minus and its parental strain do not differ in their survivability in soil and that the survival of ice minus in soil is poor. As a protective measure, a 10-meter soil barrier was designed to surround the experimental test plot to minimize the spread of ice minus beyond the application area.

There is no evidence to indicate that ice minus would colonize crop or non-commercial plants outside the test area. Ice minus was expected to occur in low numbers on plants outside the test plot, comprising only a negligible fraction of the total microbial population on plants [22]. Furthermore, any ice minus that were transported from the study area would encounter severe competition from much larger, pre-existing microbial populations. Empirical studies also indicated that ice minus was not likely to survive a full yearly cycle, since its population drops to very low levels during winter months, and its survival in soil is poor [22]. Hence, it appeared unlikely that significant numbers of ice minus would become established outside of the test plot.

5.3 Environmental Impacts

The possibility that ice minus could be a source of ice nucleation in the atmosphere and, consequently, influence precipitation patterns was also evaluated. Lindemann and Upper [24] showed that vertical fluxes of microbes from crops to the atmosphere do occur and that aerosolized organisms can be carried long distances in the environment. The theory that these airborne microbes can initiate precipitation processes has not been tested, because it is difficult to accurately identify the origin of ice nuclei in the atmosphere [9]. The question with ice minus is whether it could be transported from the test plot in sufficient numbers to cause a significant ice nucleation event. Since ice minus is not expected to thrive in competition with naturally occurring strains, it seems unlikely that it would survive insufficient numbers to affect precipitation patterns [9]. The U. S. Environmental Protection Agency consulted with several atmospheric scientists concerning the potential effects of ice minus on precipitation and also concluded that the small-scale application proposed by the University of California would not significantly influence precipitation processes [23].

5.4 Summary of Findings

The National Institutes of Health [25] reported that the probability of any adverse environmental or human health risks was at or near zero because (1) both the naturally occurring and genetically altered forms of *P. syringae* are not pathogenic to plants, animals, or humans, and (2) although small numbers of ice minus could be transported from the test area, established natural microbial populations would prevent their colonization on neighboring plants. The U. S. Environmental Protection Agency [23] also concluded that the field application of ice minus presents "no foreseeable significant risks to human health or the environment." In April, 1987, the ice minus field test was conducted without apparent adverse health or environmental impacts [26].

6. DISCUSSION OF THE ICE MINUS RISK ASSESSMENT

Assessing the level of risk associated with environmental releases of bioengineered organisms should be based on the biological and ecological characteristics of the organism from which they are derived as well as the environment into which they will be introduced. It is clear that some engineered organisms warrant greater concern than other organisms. Of particular concern are: (1) pathogenic organisms or organisms that have received genetic material from pathogens; (2) organisms not indigenous to the environment into which they will be introduced; and (3) organisms that contain genetic material from unrelated species. Moreover, any bioengineered organism that has a high potential for widespread dispersal after its introduction into the environment or the potential to establish dominance over existing populations demands detailed experimental testing and regulatory scrutiny. Ice minus is a bioengineered organism that differs from its naturally occurring parental strain through the deletion of a single gene. Extensive testing demonstrated that it is nonpathogenic to plants, animals, and humans, has a low survival potential, is not likely to be found in significant numbers in areas outside of the test site, and its behavior and fate can be predicted from knowledge of the parental strain from which it was derived. Based on the nature and extent of the scientific evidence gathered, it appears that the risk assessment methodology prepared for the proposed environmental release of ice minus was appropriate and comprehensive.

7. CONCLUSIONS

While the notion that bioengineered products are too dangerous to release into the environment seems unreasonable, legitimate public concerns must be acknowledged. An urgent need exists for the scientific community to provide investigators and regulators with guidelines for classifying and evaluating the risks associated with environmental introductions of genetically altered organisms. Although current risk assessment methodologies developed for toxic and hazardous chemicals can be applied to bioengineered organisms, some research questions cannot be answered by these established procedures [27]. While some individuals argue that there have been many successful deliberate environmental releases of bioengineered organisms [4], more research is needed to better document potential long-term consequences and to better quantify the variables affecting microorganism survival in the environment. However, given our current understanding of the environmental behavior of genetically altered microorganisms, the development of a suitable risk assessment framework for biotechnology based on available methodologies seems feasible.

8. REFERENCES

[1] Fiksel, J. and V. T. Covello, "The Suitability and Applicability of Risk Assessment Methods for Environmental Applications of Biotechnology," in J. Fiksel and V. T. Covello (eds.), "Biotechnology Risk Assessment," Pergamon Press, 1986, pp. 1-34.

[2] Ezzell, C., "U. S. Attitudes to Biotechnology Show Qualified Support," Nature, Vol. 327, 1987, p. 453.

[3] Kingsbury, D. T., "The Regulatory 'Coordinated Framework' for Biotechnology," Bio/Technol., Vol. 4, No. 12, 1986, pp. 1071-1073.

[4] Miller, H. I., "'Old' Biotechnology, 'New' Biotechnology and Risk Assessment: A Perspective," Presented at the NATO Advanced Research Workshop, Rome, Italy, June 6-10, 1987. (not yet published)

[5] Watson, J. D. and J. Tooze, "The DNA Story, A Documentary History of Gene Cloning," H. F. Freeman Press, 1981.

[6] National Academy of Sciences, "Introduction of Recombinant DNA-Engineered Organisms into the Environment: Key Issues," Prepared by the Committee on the

Introduction of Genetically Engineered Organisms into the Environment, Washington, DC, 1987.

[7] Jackson, D. A., "Witness Panel," Recom. DNA Technol. Bull., Vol. 7, 1984, pp. 196-198.

[8] Franklin, C. A., E. R. Nestmann, and L. Ritter, "Risk Assessment and Regulation of Genetically Engineered Products," in H. M. LeBaron, R. O. Mumma, R. C. Honeycutt, J. H. Duesing, J. F. Phillips, and M. J. Haas (eds.), "Biotechnology in Agricultural Chemistry," American Chemical Society Series No. 334, 1987, pp. 336-351.

[9] Lindemann, J., G. J. Warren, and T. V. Suslow, "Ice-Nucleating Bacteria," Science, Vol. 231, 1986, p. 536.

[10] Simberloff, D. S., "Witness Panel," Recom. DNA Technol. Bull., Vol. 7, 1984, pp. 196-198.

[11] Brill, W. J., "Safety Concerns and Genetic Engineering in Agriculture, "Science, Vol. 227, 1985, pp. 381-384.

[12] Gillett, J. W., "Risk Assessment Methodologies for Biotechnology Impact Assessment," Environ. Manage., Vol. 10, No. 4, 1986, pp. 512-532.

[13] Levy, S. B., "Human Exposure and Effects Analysis for Genetically Modified Bacteria," in J. Fiksel and V. T. Covello (eds.), "Biotechnology Risk Assessment," Pergamon Press, 1986, pp. 56-74.

[14] Cairns, J., Jr. and J. R. Pratt, "Ecological Consequence Assessment: Effects of Bioengineered Organisms," in J. Fiksel and V. T. Covello (eds.), "Biotechnology Risk Assessment," Pergamon Press, 1986, pp. 88-108.

[15] Barnthouse, L. W. and A. V. Palumbo, "Assessing the Transport and Fate of Bioengineered Microorganisms in the Environment," in J. Fikseland V. T. Covello (eds.), "Biotechnology Risk Assessment," Pergamon Press, 1986, pp. 109-128.

[16] Halvorson, H. O., D. Pramer, and M. Rogul (eds), "Engineered Organisms in the Environment: Scientific Issues," American Society for Microbiology, 1985.

[17] Gillett, J. W., A. M. Stern, M. A. Harwell, and S. A. Levin, "Executive Summary," Environ. Manage., Vol. 10, No. 4, 1986, pp. 437-440.

[18] Alexander, M., "Fate and Movement of Microorganisms in the Environment. Part 1. Survival and Growth of Bacteria, " Environ. Manage., Vol. 10, No. 4, 1986, pp. 463-469.

[19] Liang, L. N., J. L. Sinclair, C. M. Mallory, and M. Alexander, "Fate in Model Ecosystems of Microbial Species of Potential Use in Genetic Engineering," Appl. Environ. Microbiol., Vol. 44, No. 3, 1982, pp. 708-714.

[20] Andow, D. A., "Fate and Movement of Microorganisms in the Environment. Part 2. Dispersal of Microorganisms with Emphasis on Bacteria," Environ. Manage., Vol. 10, No. 4, 1986, pp. 470-487.

[21] Bovallius, A., R. Roffey, and E. Henningson, "Long-Range Transmission of Bacteria," Ann. N. Y. Acad. Sci., Vol. 353, 1980, pp. 186-200.

[22] University of California, "Ice Nucleating Minus Research Field Test: Draft Environmental Impact Report", Division of Agriculture and Natural Resources, Berkley, California, 1986. (unpublished)

[23] U. S. Environmental Protection Agency, "Hazard Evaluation Division's Final Position on Lindow Experimental Use Permit Applications," Memo to Tom Ellwanger, Head, Technical Support Section, from Fred Betz, Hazard Evaluation Division, Washington, DC, 1986. (unpublished)

[24] Lindemann, J. and C. D. Upper, "Aerial Dispersion of Epiphytic Bacteria over Bean Plants," Appl. Environ. Microbiol., Vol. 50, No. 5, 1985, p. 1227.

[25] National Institutes of Health, "Environmental Assessment and Findings of No Significant Impacts," in "Ice Nucleating Minus Research Field Test: Draft Environmental Impact Report Appendices," Division of Agriculture and Natural Resources, University of California, Berkley, 1985. (unpublished)

[26] Lindemann, J., Advanced Genetic Sciences, Personal Communication, 1987.

[27] Rissler, J. F., "Research Needs for Biotic Environmental Effects of Genetically Engineered Microorganisms," Recom. DNA Technol. Bull., Vol. 6, 1984, pp. 43-56.

A BIOLOGICAL APPROACH TO ASSESSING ENVIRONMENTAL RISKS OF ENGINEERED MICROORGANISMS

Lawrence W. Barnthouse
Environmental Sciences Division
Oak Ridge National Laboratory
P.O. Box 2008
Oak Ridge, Tennessee 37831-6036

Gary S. Sayler
Graduate Program in Ecology
The University of Tennessee
Knoxville, Tennessee 37996

Glenn W. Suter II
Environmental Sciences Division
Oak Ridge National Laboratory
P.O. Box X
Oak Ridge, Tennessee 37831-6038

Introduction

Many environmental applications of biotechnology involve deliberate release of organisms into the environment, where they must survive and multiply to perform their functions. Examples of such applications include degradation of toxic chemicals and in-situ leaching of ores. It is natural, when developing a scheme for assessing environmental risks of these microorganisms, to take as a point of departure existing schemes for assessing environmental risks of toxic contaminants. The components of such risk assessments, characterized by the National Academy of Sciences [1] as "hazard identification," "dose-response assessment," "exposure assessment," and "risk characterization", derive from a systematic examination of the physical, chemical, and toxicological phenomena underlying the risk: the emission rate of the toxicant, its dispersion in air and water, the chemical transformations occurring during transport, and the relationship between the dose to the exposed organism (usually man) and the toxicological effect. Consideration of these processes underlies both quantitative and qualitative risk assessments of radionuclide release, pesticide application, toxic chemical manufacture, and hazardous waste disposal.

Because microorganisms in the environment are subject to the same physical laws that govern contaminant transport, it is

NATO ASI Series, Vol. G18
Safety Assurance for Environmental Introductions
of Genetically-Engineered Organisms
Edited by J. Fiksel and V.T. Covello

feasible to adapt existing physical transport models for use with microorganisms rather than chemicals [2]. There are, however, fundamental differences between microorganisms and toxic chemicals that lead us to question the applicability of contaminant-oriented risk assessment schemes to microorganisms.

These differences suggest to us that a new conceptual approach to risk assessment is required for assessing environmental risks of microorganism releases, one that emphasizes the fundamentally biological nature of the phenomena involved.

A Biological Approach to Risk Assessment

A detailed look at the biological processes governing the transport and fate of microorganisms in real environments will show that it is much more fruitful to look at the process in terms of dispersal, establishment, and effects than in the conventional terms of release, transport, and effects.

For microorganisms that must survive and multiply in the environment to perform their intended function, the rate of dispersal per unit time from the site of initial release is a function of the carrying capacity of the release site rather than of the number of organisms released. This, in turn is a function of substrate abundance, physical/chemical limiting factors, and rates of mortality due to grazing by protozoa and other soil microfauna. In addition, the actual mobilization of organisms for transport depends on the biological phase partitioning of the organisms, i.e., whether they are in free suspension or immobilized on biofilms, or are bound to organic material or root nodules. Atmospheric and hydrologic processes are important influences on the dispersal of microorganisms, however, the number ultimately deposited on sites suitable for colonization will be largely dependent on the ability of the organisms to survive during transport. Resistance to desiccation, ability to form spores, and other biological determinants of survival during transport vary greatly among microorganisms, and the survival rate of any one type of microorganism will vary among environmental pathways [2].

Organisms deposited on sites unsuitable for survival and growth cannot proliferate and (unless they can form long-lived spores) cannot pose an environmental risk. Whether a given microorganism can become established at a site where it is deposited depends on its environmental tolerance ranges, the availability of suitable substrates, and how these are distributed in time and space. In addition, there is a minimum propagule size or rate of arrival of organisms below which establishment is unlikely because the arriving organisms die before they can reproduce. Knowledge of minimum propagule size is critical because if it is small enough (e.g., 10-20 cells), then migrating birds and insects can carry enough organisms to establish new colonies.

Much has already been written about the potential ecological effects of engineered organisms [3, 4, 5]. For the types of organisms discussed in this paper (i.e., nonpathogenic soil-dwelling microorganisms), changes in the structure and function of the microbial community are the obvious effects of concern. It is an open question whether changes in diversity or species composition are consequential endpoints for risk assessment, in and of themselves. Should we be concerned if one strain of pseudomonad is replaced by another ecologically equivalent strain? Clearly, however, the function of microorganisms is important. They have critical roles in decomposition of organic matter and in the cycling of sulfur, nitrogen, phosphorus, and other minerals in the soil. Although no examples are yet known, some kinds of engineered microorganisms (e.g., an organism deliberately designed to increase the availability of dissolved nutrients to plant roots) might alter these processes. This possibility was acknowledged in the Coordinated Framework for Regulation of Biotechnology [6].

Clearly, a quantitative risk assessment that focused on estimating the physical dispersion of microorganisms as a function of the number of organisms deposited at the site of release would ignore most of the critical processes governing the dispersal and establishment of microorganisms. A monitoring program that emphasized detecting atmospheric transport of an introduced strain would be inadequate if the organism being monitored were efficiently dispersed by birds and insects. The same program would be irrelevant if the organism were so specialized that it could not become established outside the release site, so fragile that it could not survive transport, or so similar to indigenous microorganisms that its establishment would have no observable ecological consequences.

We believe that adequate risk assessments for engineered microorganisms must be based on a conceptual scheme that emphasizes (1) dispersal, (2) establishment, and (3) effects on the structure and function of the indigenous microbial community. Adopting such a scheme would involve a significant reorientation in both the types of models that may be useful and the types of experimental data that are needed for risk assessment. The remainder of this paper discusses the kinds of models and experimental techniques that are relevant to a biologically based risk assessment scheme.

Models and Experimental Systems Suitable for Biologically Based Risk Assessments

Many suitable models and experimental systems for risk assessment of bioengineered microorganisms are available, although most have not to our knowledge been used for this purpose. The substantial body of theory on biogeography provides useful insights into the process of establishment of new colonies. MacArthur and Wilson [7] showed that the probability of any single dispersing organism establishing a new colony is a function of its probability of reproducing before it dies. The probability of any propagule establishing a population can be estimated from the ratio of the instantaneous rates of birth and death for the propagule. The smaller this ratio, the fewer organisms are required to establish a population. Other models have been developed, but their conclusions are qualitatively similar.

The theory suggests that it may be possible to estimate the minimum propagule size required to establish a new colony by measuring the rates of cell division and cell death of a strain under biologically realistic conditions. Classical closed and flow-through (chemostat) systems can provide some of this information. They can, for example, provide estimates of cell division rates and of the relationship of those rates to physicochemical tolerance limits and substrate abundance. However, classical techniques cannot address the influence of physical substrate on the growth of the organisms. In addition, pure culture systems provide little insight into death rates because in nature grazing by protozoans and other microfauna are the greatest source of mortality.

All of these can, however, be investigated in microcosm systems. Three types that appear useful for this purpose include sieved soil systems [8], intact soil cores [9], and assembled ecosystems [10, 11]. The sieved soil systems are readily reproducible and cheap, and contain both the microflora that would compete with an invading organism and the microfauna that would graze it. The intact core systems contain more realistic physical structure, and in addition contain larger soil invertebrates and plants. The assembled ecosystems are similar, but more reproducible and standardized.

These systems have the capability to capture the ecological interactions influencing birth and death, but require advances in microbiological detection techniques to be truly useful. There are such techniques. It is possible to separately measure net population growth, and reproduction in complex communities. The fluorescent antibody direct viable cell count [12] can be used to directly assay the abundance of an organism. When tracked in time, this provides a measure of the net rate of population growth. The rate of cell replication (always larger than the net growth rate) can, at least in principle, be measured using the technique of nonreplicating genetic markers [13]. The death rate can then be estimated by subtraction.

As noted by Wiegert [14], effects of engineered microorganisms on the structure and function of soil microbial communities can be approached through classical systems ecology by (1) identifying microbial types, or guilds, participating in the cycles of interest, (2) identifying linkages between pools and organisms responsible for transformations, and (3) developing and validating quantitative functions for the relationships: uptake and transformation kinetics, effects of soil conditions, and ultimately, relationships between mineral fluxes and plant production. The only models of this type of which we are aware are those of Smith [15] for the carbon and phosphorus cycles in soil.

The principal barrier to the development of microbial systems models is inadequate understanding of soil microbial communities. Microcosm systems appear, again, to be critical for making significant progress. New techniques for detecting and measuring the activities of specific microbial strains have the potential to make these systems powerful tools for advancing microbial ecology. We have already mentioned techniques such as the fluorescent antibody method [12] for detecting individual strains. Marker genes that are activated only when an organism is metabolically active can enable measurement of substrate transformation by a single target strain out of hundreds or thousands present in the soil. Finally, and perhaps most important, the physical and temporal scales of microbial activity are ideal for microcosm study. In contrast to studies of trees or fish, the microcosm can contain most of the ecosystem components that directly affect microbial activity.

To this point we have considered only organisms modified by engineering of the bacterial chromosome. These are the most straightforward to address. However, many genes of interest, such as the halogenated hydrocarbon-degrading gene described by Sayler et al [16], are carried on plasmids. Plasmid-carried genes are a serious problem for risk assessment, because many aspects of their proliferation, transport, invasion, and establishment are predicted not by the initial host organism that is introduced, but by the properties of organisms to which the plasmid may be transferred. Cell-oriented models and experimental methods are of at best limited value for detecting and modeling the activity of a gene that is transferred to many hosts. Models of the transfer of plasmids between two species exist [17]. However, the array of potential hosts in real ecosystems is far to great for explicit modeling of plasmid transfer among all possible hosts.

We believe that useful risk assessments are still possible, but they must use the plasmid rather than the host as the unit of study. Measurement techniques would have to focus on plasmid DNA [18] or on the actual gene product rather than on the host cell. Microcosms containing a natural array of potential hosts are the minimum experimental unit. Using such systems, it is feasible to estimate the size and activity of the source population of plasmids. Conservative assumptions about the emigration and establishment of the plasmid-carrying hosts might then be developed based on biological characteristics of highly mobile bacteria.

Discussion

It would rarely be necessary to quantify all aspects of the dispersal, establishment, and effects of a microbe before introduction. The primary purpose of the scheme outlined in this paper is not to provide a recipe for quantifying risks but to provide guidance for all levels of risk assessment from the qualitative to the quantitative. Our scheme could, first of all, be used to ensure that lists of criteria to be considered in approving releases of microorganisms include all of the important characteristics influencing the environmental risk of an organism. At a higher level, standardized testing requirements and decision criteria could be defined based on characteristics associated with high potential mobility, broad environmental tolerance, or involvement in critical mineral cycles.

It is easy to envision the evolution of a tiered testing/assessment scheme for microorganisms that would closely parallel the schemes now used to regulate pesticides and toxic chemicals. The lowest tier would consist of a qualitative evaluation of basic strain characteristics; higher tiers would include microcosms of graded complexity up to greenhouses or other large contained systems. Quantitative risk assessment models would be part of the highest tier.

Such a scheme has already been described [19], and we have no doubt that within a few years tiered testing systems will be used to regulate environmental releases of engineered microorganisms. It is imperative, however, that these systems be based on a biological rather than a physicochemical concept of risk. We have tried to show in this paper that most of the tools necessary for biologically based risk assessments either already exist or are within reach. Microbial ecologists and environmental regulators must now work together to integrate them into an effective regulatory system.

Acknowledgements

The authors thank F.E. Sharples and A.V. Palumbo for their reviews of this article. Preparation of this article was funded by the Oak Ridge National Laboratory Exploratory Research Program. Oak Ridge National Laboratory is operated by Martin Marietta Energy Systems, Inc. under Contract No. DE-AC05-84OR21400 with the U.S. Department of Energy. Environmental Sciences Division Publication No. 3145.

References

[1] National Academy of Sciences, "Risk Assessment in the Federal Government: Managing the Process," National Academy Press, Washington, D.C., 1983

[2] Barnthouse, L. W. and A. V. Palumbo, Assessing the transport and fate of bioengineered microorganisms in the environment, in J. Fiksel and V. Covello (eds.), "Biotechnology Risk Assessment," Pergamon Press, 1986, pp. 109-128.

[3] Sharples, F.E., Spread of organisms with novel genotypes: thoughts from an ecological perspective, Recomb. DNA Tech. Bull., Vol 6, 1983, pp. 43-56.

[4] Cairns, J., and J.R. Pratt, Ecological consequence assessment: effects of bioengineered organisms, in J. Fiksel and V. Covello (eds.), "Biotechnology Risk Assessment," Pergamon Press, 1986, pp. 88-108.

[5] Colwell, R.K., Ecology and biotechnology: expectations and outliers, in J. Fiksel and V. Covello (eds.), "Risk Analysis Approaches for Environmental Releases of Genetically Engineered Organisms," NATO Advanced Science Institutes Series, Volume F, Springer-Verlag, 1988 (this volume).

[6] Office of Science and Technology Policy, U.S., "Coordinated Framework for Regulation of Biotechnology," Federal Register, Vol. 51, 1986, pp. 23301-23350.

[7] MacArthur, R.H., and E.O. Wilson, "The Theory of Island Biogeography," Princeton University Press, 1968.
[8] Suter, G.W. II, and F.E. Sharples, Examination of a proposed test for effects of toxicants on soil microbial processes, in D. Liu and B.J. Dutka (eds.), "Toxicity Screening Procedures Using Bacterial System," Marcel Dekker, Inc., 1984, pp. 327-344.
[9] Van Voris, P., R.V. O'Neill, W.R. Emanuel, and H.H. Shugart, Functional complexity and ecosystem stability, Ecology, Vol. 6, 1980, pp. 1352-1360.
[10] Metcalf, R.L., Model ecosystem studies of bioconcentration and biodegradation of pesticides, Environ. Sci. Res., Vol. 10, 1977, pp. 127-144.
[11] Gillett, J.W., and J.D. Cole, Pesticide fate in terrestrial laboratory ecosystems, Int. J. Environ. Stud., Vol. 10, 1976, pp. 15-22.
[12] Stanley, P.M., M.A. Gage, and E.L. Schmidt, Enumeration of specific populations by immunofluorescence, in J.W. Costerton and R.R. Colwell (eds.), "Native Aquatic Bacteria: Enumeration, Activity, and Ecology," American Society for Testing and Materials, 1979, pp. 46-55.
[13] Meynell, G.G., Use of superinfecting phage for estimating the division rate of lysogenic bacteria in infected animals, J. Gen. Microbiol., Vol. 21, 1959, pp. 421-437.
[14] Wiegert, R.G., Ecosystem structural and functional analysis, in J. Fiksel and V. Covello (eds.), "Biotechnology Risk Assessment," Pergamon Press, 1986, pp. 129-143.
[15] Smith, O.L., "Soil Microbiology: A Model of Decomposition and Nutrient Cycling," CRC Press, 1982.
[16] Sayler, G.S., H.-L Kong, and M.S. Shields, Plasmid-mediated biodegradative fate of monohalogenated biphenyls in facultatively anaerobic sediments, in G.S. Omen and A. Hollaender (eds.), "Genetic Control of Environmental Pollutants," Plenum Press, 1984, pp. 117-135.
[17] Levin, B.R., and V.A. Rice, The kinetics and transfer of nonconjugative plasmids by mobilizing conjugative factors, Genet. Res. Camb., Vol. 35, 1980, pp. 241-259.
[18] Sayler, G.S., M.S. Shields, E.T. Tedford, A. Breen, S.W. Hooper, K.M. Sorotkin, and J.W. Davis, Application of DNA-DNA colony hybridization to the detection of catabolic genotypes in environmental samples, Appl. Environ. Microbiol., Vol. 49, 1985, pp. 1295-1303.
[19] Gillette, J.W., Risk assessment methodologies for biotechnology impact assessment, Environ. Manage., Vol. 10, 1986, pp. 515-532.

INFORMING PEOPLE ABOUT THE RISKS OF BIOTECHNOLOGY: A REVIEW OF OBSTACLES TO PUBLIC UNDERSTANDING AND EFFECTIVE RISK COMMUNICATION

Vincent T. Covello, Ph.D.
Center for Risk Communication
School of Public Health
Columbia University
60 Haven Ave., B-1
New York, New York 10027

Patricia Anderson, Ph.D.
National Science Foundation
1800 G St., NW
Washington, D.C. 20550

Abstract

This paper reviews the literature on informing people about the risks of biotechnology. The paper focuses on the risks of environmental releases of genetically engineered organisms. The paper describes the principle obstacles to public understanding of biotechnology risks and concludes with a set of guidelines for effective risk communication.

I. Introduction

The goal of informing people about the risks of biotechnology seems easy in principle but surprisingly difficult in practice. To be effective, officials in government and industry must overcome a number of significant obstacles. These obstacles can be organized into four conceptually distinct, but related categories: (1) characteristics and limitations of scientific data about the risks of biotechnology and environmental introductions of genetically engineered organisms; (2) characteristics and limitations of government and industry officials in communicating with the public about the risks of biotechnology; (3) characteristics and limitations of the media in reporting information about the risks of biotechnology; and (4) characteristics and limitations of the public in assimilating and understanding information about the risks of biotechnology. Each of these obstacles is discussed below.

NATO ASI Series, Vol. G18
Safety Assurance for Environmental Introductions of Genetically-Engineered Organisms
Edited by J. Fiksel and V. T. Covello

II. Characteristics and Limitations of Scientific Data About the Risks of Biotechnology

One of the principal strengths of biotechnology risk assessments is that they attempt to minimize ambiguities by providing results in the precise language of numbers. Because biotechnology risk assessments are based on the concept of decomposing a situation into its logical pieces, they also provide an effective means for organizing and analyzing complex data on possible health and environmental effects.

Despite these strengths, even the best biotechnology risk assessment cannot provide exact answers (e.g., Alexander, 1985; Davis, 1985; 1987; Colwell et al., 1985; Conservation Foundation, 1985; Fiksel and Covello, 1986; National Academy of Sciences, 1987; U.S. Environmental Protection Agency, 1987). Due to limitations in scientific understanding, data, models, and methods, the results of most biotechnology risk assessments are at best approximations. Moreover, uncertainties in the biotechnology risk assessment process often lead to different estimates of risk. For example, some have argued that the risks of planned introductions of genetically engineered organisms into the environment are no greater than the risks from other accepted activities, such as plant hybridization (e.g., Fiksel and Covello, 1986; Brill, 1985; 1988; National Academy of Sciences, 1987). However, others have argued that the risks are possibly much greater, citing the histories of imported species such as the gypsy moth, kudzu vine, Japanese beetle, and water hyacinth (e.g., Gore, 1984; Regal, 1987). Part of the criticism derives from a concern that the public will be misled by biotechnology risk assessments that claim greater precision that can reasonably be justified by the quality of the data or by the current degree of scientific understanding. These uncertainties invariably affect communications with the public in the adverserial climate that surrounds most biotechnology and genetic engineering issues.

III. Characteristics and Limitations of Government Officials, Industry Officials, and Other Sources of Information about the Risks of Biotechnology

Two of the principle sources of information about the risks of biotechnology--government agencies and industry--often lack trust and credibility. In the United States, for example, overall public confidence and trust in government and industry has declined precipitously over the past two decades (Harris, 1984; Ruckelshaus, 1983; 1984; 1987). More specifically one study found that 55 percent of the public had a great deal of confidence in major business companies in 1960. By 1980, this had dropped to 19 percent.

Trust and confidence are intimately linked and can be undermined by numerous factors. In the biotechnology area, these include public perceptions (a) that government agencies responsible for managing biotechnology safety are unduly influenced by the biotechnology industry, (b) that government agencies responsible for managing biotechnology safety are inappropriately biased in favor of promoting biotechnology products or processes that significant critics believe to be unacceptably risky, (c) that biotechnology safety managers in government agencies and industry are not technically competent, (d) that the biotechnology safety programs of government agencies and industry have been mismanaged, and (e) that biotechnology safety managers in government agencies and the biotechnology industry have lied, presented half-truths, or made serious errors in the past.

Several other factors also undermine public trust and confidence in biotechnology safety managers. First, disagreements among biotechnology risk assessment experts have undermined public trust and confidence. Because of different assumptions, data, or methods, experts in biotechnology risk assessment often engage in highly visible debates and disagreements about the reliability, validity, and meaning of biotechnology risk assessment results. In many cases, equally prominent experts have taken diametrically opposed positions on biotechnology risk assessment issues (see, for example, Gore, 1984; Brill, 1988; Young and Miller, 1987). While such debates may be constructive for the development of scientific knowledge, they often undermine public trust and confidence in government and industry officials.

A second factor undermining public trust and confidence in biotechnology safety mangers is the lack of resources for biotechnology risk assessment and management. These resources are seldom adequate to meet demands by citizens and public interest groups for definitive findings and action. Explanations by officials that the generation of health and environmental data about the risks of biotechnology and genetic engineering is expensive and time consuming--or that biotechnology risk assessment and management activities are constrained by resource, technical, statutory, legal, or other limitations--are seldom perceived to be satisfactory. Individuals facing what they believe is a new and significant health and environmental risk--such as from the field testing a new biotechnology product--are especially reluctant to accept such claims.

A third factor undermining public trust and confidence in biotechnology safety managers is the lack of adequate coordination among responsible authorities. Approaches to biotechnology risk assessment and management by government agencies and authorities at all levels--local, state, regional, national, and international--are often inconsistent (e.g., National Academy of Sciences, 1987; Brill, 1988). With rare exceptions, few requirements exist for regulatory agencies to develop coherent, coordinated, consistent, and interrelated plans, programs, and guidelines for managing biotechnology safety. As a result, regulatory systems tend to be highly fragmented. This fragmentation often leads to jurisdictional conflicts about which agency and which level of government has the ultimate responsibility for assessing and managing a new biotechnology method, product, or process. Lack of coordination, different mandates, and confusion about responsibility and authority also lead, in many cases, to the production of multiple and competing biotechnology risk assessments--each of which might provide a different estimate of risk. A commonly observed result of such confusion is the erosion of public trust, confidence, and acceptance.

A fourth factor undermining public trust and confidence in biotechnology safety managers is the lack of adequate risk communication skills. Most biotechnology safety managers lack adequate training in presentation and interaction skills (see, e.g., Covello et al., 1988a; Covello et al., 1988a; Covello et al., 1988b; Hance et al., 1987). Moreover, biotechnology safety mangers often use complex and difficult language and jargon in communicating the results of biotechnology risk assessments to the media and the public. The use of technical language or jargon is not only difficult to comprehend but can also create a perception that the expert is being unresponsive and evasive. Exacerbating the problem is the lack of attention paid to translating unfamiliar biotechnology risk assessment concepts and terms such as gene exchange, genetic alteration, and bacterial domestication into terms that the public can understand (e.g., Covello et al., 1988a).

A fifth factor undermining public trust and confidence in biotechnology safety managers is insensitivity by managers to the information needs and concerns of the public. Biotechnology safety managers often operate on the assumption that they and their audience share a common framework for evaluating and interpreting biotechnology risk assessments. However, research conducted by behavioral and social science researchers suggests that this is often not the case (e.g., Covello, 1983; Slovic, 1987; Vlek and Stallen, 1981). One of the most important findings to emerge from this literature is that people take into consideration a complex array of qualitative and quantitative factors in defining, evaluating, and acting on risk information (e.g., Slovic, et. al., 1980; Vlek and Stallen, 1981; Litai et. al., 1983; Renn, 1981; Otway and von Winterfeldt, 1982; Covello, 1983; McGuire,1981; Green, 1984a; 1984b; Janz and Becker, 1984; Rogers, 1981). For example, the following factors appear to be important in public perceptions of the risks of biotechnology and genetic engineering (Table 1).

(1) catastrophic potential, i.e., people are more concerned about adverse ecological or health effects that are grouped in time and space (e.g., deaths from airplane crashes; large scale ecological damage from a toxic spill or accidental release) than about fatalities and injuries that are scattered or random in time and space (e.g., automobile accidents);

(2) familiarity, i.e., people are more concerned about risks that they perceive to be unfamiliar (e.g., the risks of nuclear power; risks from Recombinant DNA-engineered organisms) than about risks that they perceive to be familiar (e.g., household accidents);

(3) understanding, i.e., people are more concerned about

activities characterized by poorly understood biological or ecological mechanisms or processes (e.g., the fate and transport of toxic chemicals or genetically engineered microorganisms) than about activities characterized by apparently well understood exposure mechanisms or processes (e.g., pedestrian accidents or slipping on ice);

(4) uncertainty, i.e., people are more concerned about risks that they perceive to be scientifically unknown or uncertain (e.g. the risks of nuclear power or genetic engineering) than about risks that they perceive to be relatively known to science (e.g., actuarial data on automobile accidents);

(5) controllability, i.e., people are more concerned about risks that they perceive to be not under their personal control (e.g., accidents at nuclear power stations and at biotechnology testing facilities) than about risks that they perceive to be under their personal control (e.g., driving an automobile or riding a bicycle);

(6) volition, i.e., people are more concerned about risks that they perceive to be involuntary (e.g., exposure to adverse health effects or environmental injury from industrial emissions or from deliberate releases of genetically engineered microorganisms) than about risks that they perceive to be voluntary (e.g. smoking, driving, sunbathing, or mountain climbing);

(7) effects on children, i.e., people are more concerned about activities that put children specifically at risk (e.g., milk contaminated with toxic substances or exposures to toxic substances by pregnant women) than about activities that do not put children specifically at risk (e.g., adult smoking);

(8) effects on future generations, i.e., people are more concerned about activities that they perceive pose risks to future generations (e.g., long term climatic risks or adverse ecological effects due to emissions of pollutants or from the release of genetically engineered microorganisms) than to risks that pose no special risks to future generations (e.g., skiing accidents);

(9) victim identity, i.e., people are more concerned about risks to identifiable victims (e.g., a laboratory worker unintentionally exposed to a toxic substance, a sailor lost at sea, or a child who has fallen in an abandoned well) than about risks to statistical victims (e.g., statistical profiles of automobile accident victims);

(10) dread, i.e., people are more concerned about risks that are dreaded and evoke a response of fear, terror, or anxiety (e.g., the risk of cancer or the risk of uncontrolled genetic mutations due to the release of genetically engineered microorganisms) than to risks that are not especially dreaded and do not evoke a special response of fear, terror, or anxiety (e.g., common colds and household accidents);

(11) trust in institutions, e.g., people are more concerned about situations where the responsible risk management institution is perceived to lack trust and credibility (e.g., regulatory agencies that are perceived to have inappropriately close ties to the industries they regulate) than they are about situations where the responsible risk management institution is perceived to be trustworthy and credible (e.g., trust in the Centers for Disease Control).

(12) media attention, i.e., people are more concerned about risks that receive much media attention (e.g., media coverage of debates among experts about the risks of nuclear power and biotechnology) than about risks that receive little media attention (e.g., on-the-job accidents);

(13) accident history, i.e., people are more concerned about activities that have a history of major and sometimes minor accidents (e.g., nuclear power plant accidents such as the accident at Three Mile Island and Chernobyl) than about activities that have little or no history of major or minor accidents (e.g., recombinant DNA experimentation);

(14) equity and fairness, i.e., people are more concerned about activities that are characterized by a perceived inequitable or unfair distribution of risks and benefits (e.g., the siting of biotechnology testing facilities or the siting of the first U.S. repository for high level nuclear waste) than about activities characterized by a perceived equitable or fair distribution or risks and benefits (e.g., vaccination);

(15) benefits, i.e., people are more concerned about hazardous activities that are perceived to have unclear or questionable benefits (e.g., the generation of electricity using nuclear power in a nation rich in other sources of energy) than about hazardous activities that are perceived to have clear benefits (automobile driving);

(16) reversibility,i.e., people are more concerned about activities characterized by potentially irreversible adverse effects (e.g., acid rain or ozone depletion) than about activities characterized by reversible adverse effects (e.g., injuries from sports or household accidents);

(17) personal stake, i.e., people are more concerned about activities that they believe place them (or their families) personally and directly at risk (e.g., living near a biotechnology testing facility, a hazardous industrial facility, or a toxic waste site) than about activities that do not place them (or their families) personally and directly at risk (e.g., hazardous facilities located in other remote sites);

(18) evidence, i.e., people are more concerned about risks that are based on evidence from field studies (e.g., epidemiological studies of occupational exposures) than about risks that are

based on evidence from modeling or laboratory studies (e.g., laboratory studies of toxic chemicals using mice or rats);

(19) origin, i.e., people are more concerned about risks caused by human actions and failures (e.g., industrial accidents caused by negligence, inadequate safeguards, or operator error) than about risks caused by acts of nature or God (e.g., hurricanes and exposure to geological radon).

Most of these factors imply that many members of the public will perceive the risks of biotechnology to be unacceptably high. However, other factors can be expected to diminish public perceptions of risk. For example, the perceived benefits of biotechnology function as an important counterweight to the perceived risks. Two other countervailing factors are (1) the high credibility of the National Institutes of Health, which until recently was the major regulatory agency; and (2) the excellent safety record of the biotechnology industry. Profound changes in public perceptions of risk can be expected if significant events--such as a major accident--take place related to any of these risk perception factors.

IV. Characteristics and Limitations of the Media in Reporting Information About the Risks of Biotechnology

The media play a critical role in transmitting information about technological risks, including the risks of biotechnology (e.g., Goodell, 1979; Pfund and L. Hofstadter, 1981;. Mazur, 1981; Burger, 1984; Nelkin, 1984; Klaidman, 1985). However, the media have been criticized for a variety of limitations and deficiencies. For example, the media have criticized for selective and biased reporting that tends to emphasize drama, conflict, expert disagreements, and uncertainties (e.g., Goodell, 1979; Pfund and Hofstadter, 1981; Mazur, 1981; Burger, 1984; Klaidman, 1985). The media are especially biased toward stories that contain dramatic or sensational material, such as a minor or major accident at a nuclear power plant or a biotechnology testing facility (e.g., President's Commission on the Accident at Three Mile Island, 1979). Much less attention is given to daily occurrences that kill or injure far more people each year but take only one life at a time. In reporting about health and environmental risks, journalists focus on the same concerns as the public, e.g.,. potentially catastrophic effects, lack of familiarity and understanding, involuntariness, scientific uncertainty, risks to future generations, unclear benefits, inequitable distribution of risks and benefits, and potentially irreversible effects (Combs and Slovic, 1979; Fischhoff et al., 1981).

The media have also been criticized for oversimplifications, distortions, and inaccuracies in reporting information about technological risks, including the risks of biotechnology. Studies of media reporting of health and environmental risks have documented a great deal of misinformation (Burger, 1984; Dewitt et al., 1984; Combs and Slovic, 1979; Kristiansen, 1983).

Moreover, media coverage is deficient not only in what is contained in the story but in what is left out. For example, an analysis of media reports on cancer risks from various sources (Freimuth et al.; 1984) noted that these reports were deficient in (a) providing few statistics on general cancer rates for purposes of comparison; (b) providing little information on common forms of cancer; (c) not addressing known sources of public ignorance about cancer; and (d) providing almost no information about detection, treatments, and other protective measures.

Many of these problems stem from characteristics of the media and the constraints under which reporters work (see, for example, Klaidman, 1985; Sandman, 1986; Sandman et al., 1987; Nelkin, 1986). First, most reporters work under extremely tight deadlines. Second, with few exceptions, reporters do not have enough time or space to deal adequately with the complexities and uncertainties surrounding most health and environmental issues. Third, journalists achieve objectivity in a story by balancing opposing views. Truth in journalism is different from truth in science. In journalism, there is no such thing as objective truth--or at least no way to determine such truth; there are only conflicting claims, to be covered as fairly as possible. Fourth, journalists are source dependent. Under the pressure of deadlines and other constraints, reporters tend to rely heavily on sources that are easily accessible and willing to speak out. Sources that are difficult to contact, hard to draw out, or reluctant to provide interesting and non-qualified statements--are often left out. Finally, few reporters have the scientific background or expertise needed to evaluate complex scientific data and disagreements about health and environmental risks. For example, in order to write an adequate story on the risks of biotechnology, a journalist would have to understand material from molecular biology, microbiology, ecology, agricultural science, toxicology, meteorology, economics, law, and policy science.

V. Characteristics and Limitations of the Public in Assimilating and Understanding Information about the Risks of Biotechnology

Two of the most important sources of information on characteristics of the public in assimilating and understanding information about the risks of biotechnology are public opinion surveys and social psychological studies of risk perception. The main findings of both types of studies are described below.

(1) Public Opinion Surveys

Public opinion surveys are a primary source of data on public attitudes toward biotechnology and genetic engineering. Unfortunately, public opinion survey data on public attitudes toward biotechnlogy and genetic engineering are relatively sparse. To date, only three major surveys have examined public attitudes toward biotechnology. These are (1) a national survey of adult Americans conducted by Louis Harris and Associates for the Congressional Office of Technology Assessment (Office of Technology Assessment, 1987); (2) a sample of leaders in science policy, environmental organizations, and religious leaders conducted by the Public Opinion Laboratory under the direction of Jon D. Miller (Miller, 1985); and (3) two surveys conducted in 1983 and 1985 for the National Science Foundation (NSF) on public attitudes toward science and technology which included several questions relating to genetic engineering (National Science Foundation, 1983; 1985). The principal findings from these surveys are described below.

1. Public Knowledge and Awareness of Biotechnology and Genetic Engineering.

All three surveys questioned people about their general awareness and knowledge of biotechnology and genetic engineering. The surveys found that public understanding of biotechnology and genetic engineering is poor. For example, the Office of Technology Assessment (OTA) survey found that almost two-thirds of the general public know relatively little or almost nothing about biotechnology and genetic engineering.

Although the three surveys agree on this general point, they differ on how knowledgeable people are about the vocabulary of biotechnology and genetic engineering. The OTA data indicate, for example, that about half of the public thinks it understands the meaning of the term "DNA". In the 1985 NSF survey, however, only 14 percent of the general public claimed to have a clear understanding of the term "DNA".

2. Attitudes toward Risks and Benefits.

All the surveys contain questions about people's attitudes towards the risks of biotechnology and genetic engineering. In Miller's leadership survey, for example, leaders were concerned about the possibility of scientists creating undesirable organisms. In the OTA survey, only about 20 percent of the general public stated that they were aware of potential dangers associated with genetically engineered products. But of this 20 percent, less than two-thirds could name a specific danger. When respondents in the survey were asked how likely it is that genetically engineered products might pose a serious danger, half of the respondents thought it was very likely or somewhat likely. However, more than half of the respondents agreed that the risks of genetic engineering have been greatly exaggerated.

In response to a question on biotechnology regulation, the OTA study found that about three-quarters of the general public agreed with the statement: "The potential danger from genetically altered cells and microbes is so great that strict regulations are necessary." This finding contrasts with results from the Miller survey conducted two years earlier, which found that about two-thirds of science policy leaders and about one-half of both the environmental and religious leaders viewed the current level of government regulation of recombinant DNA research as about right. At least two reasons can be offered for this difference between the two surveys. First, the leaders and the general public may differ in their views on biotechnology regulation. Second, events that occurred during the two years between the surveys may have significantly affected public attitudes toward biotechnology risks and the need for regulation.

In addition to questions about risks, the surveys also contained questions about people's perceptions of the benefits of biotechnology and genetic engineering. In Miller's leadership survey, for example, leaders were asked to name benefits that might result from biotechnology and genetic engineering. Almost all leaders mentioned at least one medical benefit. Nearly all the leaders were also broadly aware of the potential for advancements in agriculture.

The NSF and Miller surveys also asked respondents to weigh the benefits against the risks of biotechnology and genetic engineering. In the 1985 NSF survey, for example, 50 percent of the respondents stated that the benefits of genetic engineering are greater than the risks. In Miller's survey, a majority of the respondents in all three leadership groups felt that the benefits of recombinant DNA technology were either slightly greater or much greater than its risks.

In both the NSF and OTA surveys, people were asked if biotechnology would make the quality of life better or worse. In the NSF survey, about two-thirds of the public stated that genetic engineering would make life better; only 16 percent felt it would make life worse. This contrasts sharply with the results for nuclear power, where public opinion was split almost equally. Similar results were obtained for a comparable question asked in the OTA survey.

3. Attitudes Toward the Release of Genetically Engineered Organisms.

Both the Miller and OTA surveys asked people about their attitudes toward the deliberate release of genetically-engineered organisms into the environment. The Miller survey asked leaders whether or not field testing of genetically-engineered organisms should be allowed or prohibited. Eighty-four percent of the science policy leaders--and about two-thirds of the environmental and religious leaders--thought such tests should be allowed. In the OTA survey, a more specific question was asked: "Do you think that environmental applications of genetically altered organisms to increase agricultural productivity or clean up environmental pollutants should be permitted on a small-scale, experimental basis, or not?" Eighty-two percent of the general public stated that such testing should be permitted. However, given a potential bias in the question introduced through the mention of the potential benefits of small-scale testing--it is not clear how to interpret this survey result.

A similar potential bias was introduced in another OTA survey question, which asked respondents whether or not they would favor a test of a genetically altered organism--such as bacteria that protects strawberries from frost --in their own community, if there were no direct risk to humans and a very remote potential risk to the local environment. In spite of the question's emphasis on benefits and low risk, only about half of the respondents were either strongly or somewhat in favor of testing this product in their community.

4. Attitudes Toward Institutions.

The OTA survey contains several questions that bear on the trust and credibility of institutions responsible for testing, regulating, or promoting genetic engineering technology. For example, when asked if commercial firms should be permitted to apply genetically engineered organisms on a large scale basis, even if the risks of environmental damage were judged to be small, a majority (53 percent) of the OTA survey respondents said they should not be permitted. Only 13 percent of the public felt that companies should be allowed to decide on the suitability of large-scale applications of genetically engineered organisms.

Respondents to the OTA survey also indicated a lack of trust in government agencies. When respondents were asked if they would believe a federal agency or a national environmental group regarding the risk attached to the use of a genetically altered organism, a majority (63 percent) said they would believe the environmental group. Only one-fourth of the public stated that they would believe the federal agency.

(2) Social Psychological Studies of Risk Perception

Another important source of information of public attitudes toward biotechnology are social psychological studies of risk perception. In recent years, a large amount of research has been conducted exploring generic characteristics and limitations of the human mind in assimilating and understanding information about health and environmental risks (see, e.g., Kasperson and Kasperson, 1983; Covello, 1983; Green and Johnson, 1983; Green, 1984a; 1984b; Becker and Maiman, 1983; Janz and Becker, 1984; McGuire, 1981; Fischhoff, 1985; Slovic, 1987). Several of the most important research findings and conclusions that bear on public perceptions of the risks of biotechnology and genetic engineering are described below.

<u>People often have inaccurate perceptions of risks.</u>

People often do not possess accurate information about specific risks. For example, researchers have found that people tend to overestimate the risks of dramatic or sensational causes of death, such as nuclear power plant accidents and homicides, and underestimate the risks of undramatic causes, such as asthma, emphysema, and diabetes, which take one life at a time and are common in nonfatal forms (Lichtenstein et al, 1978; Morgan et al., 1985). A partial explanation for this finding, is that risk judgments are significantly influenced by the memorability of past events and by the imaginability of future events (Lichtenstein et al., 1978). As a result, any factor that makes a hazard unusually memorable or imaginable, such as a recent disaster, intense media coverage, or a vivid film (e.g., the film "The Andromeda Strain") can seriously distort perceptions of risk by heightening the perception of risk. Conversely, risks that not memorable, obvious, palpable, tangible, or immediate tend to underestimated.

__People often have difficulty understanding and appreciating probabilistic information about risks.__

Researchers have found that people often have difficulty understanding and interpreting probabilistic information, especially when the risk is new and when probabilities are small. More specifically, a variety of cognitive biases and fallacies hamper people's understanding of probabilities, which in turn hamper discussions of low-probability/high consequence events and "worst case scenarios." For example, because of the difficulty people have appreciating the improbability of extreme but imaginable consequences, imaginability often blurs the distinction between what is remotely possible and what is probable (e.g., Sjoberg, 1979).

__People often respond emotionally to risk information__

People often respond emotionally to information about threats to health, safety, or the environment. Strong feeling of fear, hostility, anger, outrage, panic, and helplessness are often evoked by dreaded or newly discovered risks. Such feelings tend to be most intense when people perceive the risk to be (1) involuntary (imposed on them without their consent), (2) unfair, (3) not under their control; and (4) low in benefits. More extreme emotional reactions are generated when the risk is particularly dreaded--e.g., cancer and birth defects--and when worst case scenarios are presented (Fischhoff, 1985).

__People often display a marked aversion to uncertainty in risk information.__

Research has shown that, wherever possible, people attempt to reduce the anxiety generated by uncertainty in risk information through a variety of psychological strategies (Fischhoff, 1985; Slovic, 1987). In dealing with many health and environmental issues, this aversion to uncertainty often translates into a marked preference for statements of fact over statements of probability--the language of biotechnology risk assessment. People often demand to be told exactly what will happen, not what might happen (Alfidi, 1971; Fischhoff, 1983; Weinstein, 1979).

__People tend to ignore evidence that contradicts their current beliefs about risks.__

Research has shown that strong beliefs about risks, once formed, change very slowly and are extraordinarily persistent in the face of contrary evidence (see,e.g., Nisbett and Ross, 1980). Moreover, initial beliefs about risks tend to structure the way that subsequent evidence is interpreted. New evidence--e.g., evidence produced by a radiation risk assessment--appears reliable and informative only if it is consistent with one's initial belief; contrary evidence is dismissed as unreliable, erroneous, irrelevant, or unrepresentative.

People's beliefs and opinions are easily manipulated by the way information about risks is presented when the beliefs are weakly held.

When people lack strong prior beliefs or opinions, subtle changes in the way that risks are expressed can have a major impact on perceptions, preferences, and decisions. In recent years, for example, researchers have published numerous studies demonstrating the powerful influence of presentation or "framing effects" (see, e.g., Tversky and Kahneman; 1981; Slovic et. al., 1982). Experimental studies of these effects suggest that risk communicators have considerable ability to manipulate perceptions and behavior when beliefs and opinions are not strongly held. Several studies have dramatically demonstrated this phenomenon. For example, McNeil et. al. (1982) asked a group of subjects to imagine that they had lung cancer and had to choose between two therapies, surgery or radiation. The two therapies were described in some detail. One group of subjects was then presented with information about the probabilities of surviving for varying lengths of time after the treatment. Another group of subjects received the same information but with one major difference--probabilities were expressed in terms of dying rather than surviving (i.e., instead of being told that 68 percent of those having surgery will survive after one year, they were told that 32 percent will die). Presenting the statistics in terms of dying resulted in a dramatic drop in the percentage of subjects choosing radiation therapy over surgery (from 44 percent to 18 percent). Virtually the same results were observed for a subject population of physicians as for a subject population of laypersons.

People often consider themselves personally immune to many risks.

People often ignore risk assessment information because of unrealistic optimism and overconfidence, e.g., a belief held by an individual that fate or luck is on his side and that it "can't happen to me." This is especially true for activities that require skill and that individuals believe to be under their own personal control, such as driving or skiing. Overconfidence has been found in numerous investigations, including studies of operator errors in industrial facilities and studies showing that 95 percent of all drivers think that they are better than average drivers.

People often perceive accidents as signals

Research suggests the significance of an accident is determined only in part by the deaths, injuries, or damage that occurs. Of equal, and in some cases greater importance, is what the accident signifies or portends (Slovic et al., 1984). A major accident that causes many deaths and or extensive damage may nonetheless have only minor social significance (beyond that the victims' families and friends) if the accident occurs as part of a familiar and well understood system (e.g., a train wreck). However, a minor accident in an unfamiliar facility--or in a facility that is perceived to be poorly understood, such as a nuclear reactor or a biotechnology laboratory--have major social significance if the accident is perceived to be a harbinger of future, and possibly, catastrophic events.

People often use health and environmental risks as a proxy or surrogate for other social, political, or economic issues and concerns.

Research on the social and cultural construction and selection of risk suggests that people do not focus on particular risks simply in order to protect health, safety or the environment (Douglas and Wildavsky, 1982; Short, 1984; Johnson and Covello, 1987). The choice also reflects their beliefs about values, social institutions, nature, and moral behavior. Risks are exaggerated or minimized according to the social, cultural, and moral acceptability of the underlying activities. As a result, debates about risks are often are proxy or surrogate for debates about more general social, cultural, economic, and political issues concerns. The debate about nuclear power, for example, has often been interpreted as less a debate about the specific risks of nuclear power generation than about other fears and concerns, including the proliferation of nuclear weapons, the adverse effects of nuclear waste disposal, the value of large-scale technological progress and growth, and the centralization of political and economic power in the hands of technological elite (Covello, 1984).

One consequence of the social selection process is that risks that are finally selected for attention and concern are not necessarily chosen because of scientific evidence about their absolute or relative magnitude of possible adverse consequences. In some cases, moreover, the risks that are selected are among the least likely to affect people. In the United States, for example, the dominant risks to health are those associated with cardiovascular disease, lung cancer due to cigarette smoking, and automobiles accidents. However, in recent years Americans have focused much of their attention and resources on the risks of cancer due to industrial chemicals and radiation. This focus has persisted despite a fragile consensus among scientists that only a small fraction of all current deaths due to cancer, in the United States and elsewhere, could be due to these causes.

An important conclusion of the literature on the social and cultural construction of risk is that risk is not an objective phenomenon perceived in the same way by all interested parties.

Instead, it is a psychological and social construct, its roots deeply embedded in the workings of the human mind and in a specific social context. Each individual and group assigns a different meaning to the risk information. As in the Japanese story Roshomon, there are multiple truths, multiple ways of seeing, perceiving, and interpreting events. Each interested party--including those who generate the risk, those who attempt to manage it, those who experience it--see it in different ways.

VI. Conclusions

An appreciation of these characteristics and limitations is critical to the development of effective communication about the risks of biotechnology and genetic engineering. A start in this direction is being made in the growing literature on successful and unsuccessful cases of risk communication (see, for example, Davies, Covello, and Allen, 1987; Hance et al., 1987; Covello and Allen, 1988; Covello et al., 1987; Sandman, 1986; Sandman et al., 1987; Smith et al., 1987; Bean, 1987; Kasperson, 1986; Covello et al., 1988b). A wide variety of risk communication cases are covered by these studies. It is clear from this literature that there are no easy prescriptions for successful risk communication. However, those who have studied and participated in recent debates about risk generally agree on seven cardinal rules (Covello and Allen, 1988). Although many of these rules may seem obvious, they are continually and consistently violated in practice. Thus a useful question--for both researchers and practitioners--is why they are frequently not followed.

RULE 1. ACCEPT AND INVOLVE THE PUBIC AS A LEGITIMATE PARTNER

A basic tenet of risk communication in a democracy is that people and communities have a right to participate in decisions that affect their lives, their property, and the things they value. The goal of risk communication in a democracy should be to produce an informed public that is involved, interested, reasonable, thoughtful, solution-oriented, and collaborative; it should not be to diffuse public concerns or replace action.

Guidelines: Demonstrate your respect for the public and underscore the sincerity of your risk communication efforts by involving the community early, before important decisions are made. Make it clear that you understand the appropriateness of basing decisions about risks on factors other than the magnitude of the risk. Involve all parties that have an interest or stake in the issue.

RULE 2. PLAN CAREFULLY AND EVALUATE PERFORMANCE.

There is no such entity as "the public;" instead, there are many publics, each with its own interests, needs, concerns, priorities, preferences, and organizations. Different risk communication audiences and objectives require different risk communication strategies. Risk communication will be successful only if carefully planned.

Guidelines: Begin with clear, explicit risk communication objectives--such as providing information to the public, motivating individuals to act, stimulating response to emergencies, or contributing to the resolution of conflict. Classify and segment the various groups in your audience. Aim your communications at specific subgroups in your audience. Respect community norms about dress and behavior. Recruit spokespeople who are good at presentation and interaction. Train your staff--including technical staff--in communication skills; reward outstanding performance. Whenever possible, pretest your messages. Carefully evaluate your efforts and learn from past mistakes.

RULE 3. LISTEN TO YOUR AUDIENCE.

People in the community are often more concerned about such issues as trust, credibility, control, competence, voluntariness, fairness, caring, and compassion than about mortality statistics and the details of quantitative risk assessment. If you do not listen to people, you cannot expect them to listen to you. Communication is a two-way activity.

Guidelines: Do not make assumptions about what people know, think, or want done about risks. Take the time to find out what people are thinking: use techniques such as interviews, focus groups, and surveys. Let all parties that have an interest or stake in the issue be heard. Recognize people's emotions. Let people know that you understand what they said, addressing their concerns as well as yours. Recognize the "hidden agendas," symbolic meanings, and broader economic or political considerations that often underlie and complicate the task of risk communication.

RULE 4. BE HONEST, FRANK, AND OPEN.

In communicating risk information, trust and credibility are your most precious assets. Trust and credibility are difficult to obtain. Once lost they are almost impossible to regain completely.

Guidelines: State your credentials, but do not ask or expect to be trusted by the public. If you do not know an answer or are uncertain, say so. Get back to people with answers. Admit mistakes. Disclose risk information as soon as possible (emphasizing any reservations about reliability). If in doubt, lean toward sharing more information, not less--or people may think you are hiding something. Discuss data uncertainties, strengths and weaknesses--including the ones identified by other credible sources. Identify worst case estimates as such, and cite ranges of risk estimates when appropriate.

RULE 5. COORDINATE AND COLLABORATE WITH OTHER CREDIBLE SOURCES.

Allies can be effective in helping you communicate risk

information. Few things make risk communication more difficult than conflicts and public disagreements with other credible sources.

Guidelines: Take time to coordinate all inter-organizational and intra-organizational communications. Devote effort and resources to the slow, hard work of building bridges with other organizations. Use credible and authoritative intermediaries. Try to issue communications jointly with other trustworthy sources (credible university scientists, physicians, or trusted local officials).

RULE 6. MEET THE NEEDS OF THE MEDIA.

The media are a prime transmitter of information on risks; they play a critical role in setting agendas and in determining outcomes. The media are frequently more interested in politics than in quantitative risk statistics; more interested in simplicity than in complexity; more interested in danger than in safety.

Guidelines: Be open with and accessible to reporters. Respect their deadlines. Provide risk information tailored to the needs of each type of media (for example, graphics and other visual aids for television). Prepare in advance and provide background material on complex risk issues. Do not hesitate to follow up on stories with praise or criticism, as warranted. Try to establish long-term relationships of trust and respect with specific editors and reporters.

RULE 7. SPEAK CLEARLY AND WITH COMPASSION

Technical language and jargon are useful as professional shorthand. But they are barriers to successful risk communication. If people are sufficiently motivated, they are quite capable of understanding complex risk information, even if they may not agree with you.

Guidelines: Use simple, nontechnical language. Use vivid, concrete images that communicate on a personal level. Use examples and anecdotes that make technical risk data come alive. Avoid distant, abstract, unfeeling language about deaths, injuries, and illnesses. Acknowledge and respond (both in words and with actions) to emotions that people express--anxiety, fear, anger, outrage, helplessness. Acknowledge and respond to the distinctions that the public views as important in evaluating risks, e.g., voluntariness, controllability, familiarity, dread, origin (natural or manmade), benefits, fairness, and catastrophic potential. Use risk comparisons to help put risks in perspective; but avoid comparisons that ignore distinctions that people consider important. Always try to include a discussion of actions that are under way or can be taken. Tell people what you cannot do. Promise only what you can do, and be sure to do what you promise. Never let your efforts to inform people about risks prevent you from acknowledging--and saying--that any illness, injury, or death is a tragedy. Regardless of how well you communicate risk information, some people will not be satisfied.

Analyses of case studies suggest that these rules and guidelines form the basic building blocks for effective communication about all health and environmental risks, including the risks of biotechnology and genetic engineering (see, e.g., Davies et al., 1987; Hance et al., 1987; Covello et al., 1988a; 1988b). Each rule recognizes, in a different way, that effective risk communication is an interactive process based on mutual trust, cooperation, and respect among all parties. And each rule addresses, from a different perspective, the single, most important element in effective communication about the risks of biotechnology and genetic engineering: trust and credibility.

TABLE 1: Factors Involved in Public Risk Perception

Factor	Conditions Associated With Increased Public Concern	Conditions Associated With Decreased Public Concern
Catastrophic Potential	Fatalities and Injuries Grouped in Time and Space	Fatalities and Injuries Scattered and Random
Familiarity	Unfamiliar	Familiar
Understanding	Mechanisms or Process Not Understood	Mechanisms or Process Understood
Uncertainty	Risks Scientifically Unknown or Uncertain	Risks Known to Science
Controllability (Personal)	Uncontrollable	Controllable
Voluntariness of Exposure	Involuntary	Voluntary
Effects on Children	Children Specifically at Risk	Children Not Specifically at At Risk
Effects Manifestation	Delayed Effects	Immediate Effects
Effects on Future Generations	Risk to Future Generations	No Risk to Future Generations
Victim Identity	Identifiable Victims	Statistical Victims
Dread	Effects Dreaded	Effects Not Dreaded
Trust in Institutions	Lack of Trust in Responsible Institutions	Trust in Responsible Institutions
Media Attention	Much Media Attention	Little Media Attention
Accident History	Major and Sometimes Minor Accidents	No Major or Minor Accidents
Equity	Inequitable Distribution of Risks and Benefits	Equitable Distribution of Risks and Benefits
Benefits	Unclear Benefits	Clear Benefits
Reversibility	Effects Irreversible	Effects Reversible
Personal Stake	Individual Personally at Risk	Individual Not Personally at Risk
Scientific Evidence	Risk Estimates Based on Human Evidence	Risk Estimates Based on Animal Evidence
Origin	Caused by Human Actions or Failures	Caused by Acts of Nature or God

NOTES

This review draws heavily on the work of Paul Slovic and Detlof von Winterfeldt, to whom we are in debt. We would also like to thank the following individuals for their helpful comments on other drafts of this paper: Michael Baram, Roger Kasperson, Lester Lave, Granger Morgan, Jeryl Mumpower, Peter Sandman, Paul Slovic, and Detlof von Winterfeldt.

The views expressed in this article are solely those of the authors and do not necessarily represent the views of their organization.

REFERENCES

Alfidi, R. J.: 1971, "Informed consent: A study of patient reaction." Journal of American Medical Association, 216, 1971, 1325.

Alexander, M.: "Ecological Consequences: Reducing the Uncertainties," Issues in Science and Technology 1 (Spring 1985): 57-68.

Bean, M.: 1987, "Tools for Environmental Professionals Involved in Risk Communication at Hazardous Waste Facilities Undergoing Siting, Permitting, or Remediation." Report No. 87-30.8. Reston, Virginia: Air Pollution Control Association.

Becker, M.H. and L.A. Maiman: 1985. "Models of Health-related Behavior." Chapter 5 in D. Mechanic (ed.). Handbook of Health, Health Care, and the Health Professions. New York: The Free Press.

Brill, W.J.: 1985, "Safety Concerns and Genetic Engineering in Agriculture," Science 227 (January 25, 1985): 381-384.

Brill, W.J.: "Why Engineered Organisms Are Safe," Issues in Science and Technology, Vol. 4 (3),(Spring 1988), 44-50.

Burger, E. J. Jr.: 1984 Health Risks: The Challenge of Informing the Public. Washington, D. C.: The Media Institute, 1984.

Burton, I., Kates, R., and G. White.: 1978, The Environment as Hazard. Oxford University Press. New York.

Buss, D.M., K.H. Craik, and K.M. Dake: 1986, "Contemporary Worldviews and Perception of the Technological System," in V.

Colwell, R.K., E.A. Norse, D. Pimentel, F.E. Sharples, and D. Simberloff, "Genetic Engineering in Agriculture," letter to the editor, Science 229 (July 12, 1985): 111-112.

Journalism Quarterly, 56, 1979, 837-843.

Conservation Foundation.:1985, Risk Assessment and Risk Control, Washington, D.C.: Conservation Foundation.

Council on Environmental Quality: 1984, Environmental Quality: Annual Report, Washington, D.C.: Council on Environmental Quality.

Covello, V. T.:1983, "The perception of technological risks: A literature review". Technological Forecasting and Social Change, 23, 285-297.

Covello, V. T. Uses of social and behavioral research on risk.Environment International, 1984, 10, June, 541-545.

Covello, J. Menkes, and J. Mumpower (eds.), Risk Evaluation and Management, New York: Plenum, 1986.

Covello, V.T., von Winterfeldt, D., and Slovic, P., Communicating scientific information about health and environmental risks: problems and opportunities from a social and behavioral perspective. In V. Covello, A. Moghissi, and V.R.R. Uppuluri (eds.) Uncertainties in Risk Assessment and Risk Management. New York: Plenum Press. 1987.

Covello, V.T., D. von Winterfeldt, and P. Slovic: 1988, Communicating Risk Information to the Public, New York: Cambridge University Press (in press)

Covello, V. and F. Allen (1988), "Seven Cardinal Rules of Risk Communication," Washington, D.C.: U.S. Environmental Protection Agency, Office of Policy Analysis.

Covello, V., D. von Winterfeldt, and P. Slovic (1987), Communicating Risk Information to the Public, in C. Davies, V. Covello, and F. Allen, Risk Communication: Washington, D.C.: Conservation Foundation.

Covello, V., D. McCallum, and M. Pavlova, eds.: 1988a, Effective Risk Communication: The Role and Responsibility of Government. New York: Plenum.

Covello, V., P. Sandman, and P. Slovic: 1988b, Risk Communication, Risk Statistics, and Risk Comparisons: A Manual for Plant Managers, Washington, D.C.: Chemical Manufacturers Association.

Craik, K. H.: 1985, "Psychological Perspectives on Technology as Societal Option, Source of Hazard, and Generator of Environmental Impacts. In V.T. Covello, J.L. Mumpower, P.J.M. Stallen, and V.R.R. Uppuluri (eds.), Technology Assessment, Environmental Impact Assessment and Risk Analysis: Contributions from the Psychological and Decision Sciences. New York: Springer-Verlag, 1985.

Crouch, E., and R. Wilson: 1982, Risk/Benefit Analysis, Ballinger, Publishing Company, Cambridge, Massachusetts.

Davis, B.D., "Is Deliberate Introduction Ecologically Any More Threatening than Accidental Release?" Genetic Engineering News (October 1987: 4.

Davies, J.C., V.T. Covello, and F.W. Allen (eds.): 1987, Risk Communication, Washington, D.C.: The Conservation Foundation.

Dinman, B.D.: 1980, "The Reality and Acceptance of Risk," Journal of the American Medical Association, Vol. 244(11):1126-1128.

Dixon, B: 1985, Engineered Organisms in the Environment: Scientific Issues. Washington, D.C.: American Society for Microbiology.

Doll, R. and R. Peto: 1981, "The Causes of Cancer: Quantitative Estimates of Avoidable Risks of Cancer in the U.S. Today." Journal of the National Cancer Institute. Vol. 66, pp. 1191-1308.

Douglas, M. and A. Wildavsky: 1982, Risk and Culture. Berkeley: University of California Press.

Fiksel, J. and V. Covello: 1986, Biotechnology Risk Assessment. New York: Pergamon.

Fischhoff,B. Informed Consent for Transient Nuclear Workers. In R. Kasperson and R. W. Kates (Eds.), Equity Issues in Radioactive Waste Disposal. Cambridge, MA: Oelgeschlager, Gunn and Hain, 1983.

Fischhoff, B.: 1985, "Managing Risk Perception," Issues in Science and Technology 2(1),83-96.

Fischhoff, B.: 1985, "Protocols for Environmental Reporting: What to Ask the Experts. The Journalist, Winter, 1985, pp.11-15.

Fischhoff, B., Watson, S., and Hope, C.: 1984, "Defining risk". Policy Sciences, 17, 123-139.

Fischhoff, B., P. Slovic, and S. Lichtenstein: 1979,"Weighing the Risks," Environment 21 (4), 17-20 and 32-38.

Fischhoff, B., P. Slovic, and S. Lichtenstein: 1983, "The Public Versus the Experts: Perceived Versus Imagined Sources of Disagreements about Risks," in V. Covello, G. Flamm, J. Rodricks, and R. Tardiff (eds.), The Analysis of Actual Versus Perceived Risks, New York: Plenum. 1983.

Fischhoff, B., S. Lichtenstein, P. Slovic, S. Derby, and R. Keeney: 1981, Acceptable Risk, New York: Cambridge University Press.

Freimuth, V. S., Greenberg, R. H., DeWitt, J., and Romano, R.: 1984, "Covering cancer: Newspapers and the public interest." Journal of Communications, 34, 1984, 62-73.

R. Goodell: 1979, "The Gene Craze," Columbia Journalism Review 17, 46-51 (1979).

Gore, A., Jr.: 1984, The Environmental Implications of Genetic Engineering: Staff Report Prepared for the Subcommittee on Investigations and Oversight for the Committee on Science and Technology, Washington, D.C.: U.S. House of Representatives.

Green, L.W.: 1984a. "Health Education Models." in J.D. Matarzarro, N.E. Miller, S.M. Weiss (eds.), Behavioral Health: A Handbook of Health Enhancement and Disease Prevention. Silver Spring, MD.: John Wiley.

Green, L.W.: 1984b, "Modifying and Developing Health Behavior." Annual Review of Public Health 5: 215-236.

Green, L.W. and K.W. Johnson: 1983. "Health Education and Health Promotion." Chapter 33 in D. Mechanic (ed.). Handbook of Health, Health Care, and the Health Professions. New York: The Free Press.

Hance, B., C. Chess, and P. Sandman: 1987, Improving Dialogue with Communities: A Risk Communication Manual for Government. Trenton, New Jersey, Office of Science and Research, New Jersey Department of Environmental Protection, December, 1987

Handler, P.: 1979, "Some Comments on Risk Assessment," in National Research Council, Current Issues and Studies, Annual Report. Washington, D.C.: National Academy of Sciences.

Hohenemser, C., R.W. Kates, and P. Slovic: 1983, "The Nature of Technological Hazard," Science 220, 378-384.

Ibrekk, H. and Morgan, M.G.: 1987, "Graphical Communication of Uncertain Quantities to Non-Technical People," Risk Analysis (in press)

Janz, N.K. and Becker, M.H.: 1984: "The Health Belief Model: A Decade Later." Health Education Quarterly 11:1-48.

Johnson, B. and Covello, V.T. (eds.): 1987, The Social and Cultural Construction of Risk. Boston: Reidel.

Kahneman D. and A. Tversky: 1984. "Choices, Values, and Frames," American Psychologist, April 1984, 39:4., 341-350.

Kahneman, D., P. Slovic, and A. Tversky (eds.): 1982, Judgment Under Uncertainty: Heuristics and Biases, New York: Cambridge University Press.

Kasperson, R.: 1986, "Six Propositions on Public Participation and Their Relevance to Risk Communication." Risk Analysis, 1986, 6, 275-282.

Kasperson, R. and Kasperson, J.: 1983, "Determining the Acceptability of Risk: Ethical and Policy Issues," in J. Rogers and D. Bates, (eds.), Risk: A Symposium, Ottawa: The Royal Society of Canada, 1983.

Klaidman, S.: 1985, "Health Risk Reporting," Washington, D.C.: Institute for Health Policy Analysis, Georgetown University.

Kristiansen, C. M.: 1983, "Newspaper coverage of diseases and actual mortality statistics." European Journal of Social Psychology, 13, 1983, 193-194.

Kunreuther, H., Ginsberg, R. Miller, L., Sagi, P., Slovic, P., Borkan, B., and Katz, N.: 1978, Disaster Insurance Protection: Public Policy Lessons. New York: Wiley.

Larkin, J. and H. Simon: 1987, "Why is a Diagram (Sometimes) Worth a Thousand Words," Cognitive Science, Vol. 11: 65-99.

Lave, L.: 1981, The Strategy of Social Regulation. Washington, D.C., The Brookings Institution, 1981.

Lave, L: 1984, "Regulating Risks," Risk Analysis 4, 79-81.

Lave, L.B.: 1987, "Health and Safety Risk Analyses: Information for Better Decisions," Science, Vol. 236 (17 April 1987), 291-295.

Lichtenstein, S., Slovic, P., Fischhoff, B., Layman, M., and Coombs, B.: 1978. "Judged frequency of lethal events". Journal of Experimental Psychology: Human Learning and Memory, 4, 551-578.

Litai, D., D. Lanning, and N.C. Rasmussen: 1983, "The public perception of risk," in The Analysis of Actual Versus Perceived Risks, V.T. Covello, W. Flamm, J. Rodricks, and R. Tardiff (eds.), pp. 213-233. New York: Plenum.

Lowrance, W. W.: 1976 Of Acceptable Risk: Science and the Determination of Safety. Los Altos, CA: Kaufman, 1976.

Lowrance, W. W.: 1983. "The Agenda for Risk Decisionmaking," Environment 25(10): 4-8.

Mazur, A.: 1981, "Media Coverage and Public Opinion on Scientific Controversies," Journal of Communication, pp. 106-115.

McGuire, W.: 1981, "Theoretical Foundations of Public Communication Campaigns," pp. 41-70 in in R.E. Rice and W.J. Paisley (eds.), Public Communication Campaigns. Beverly Hills: Sage Publications.

McNeil, B.J., Pauker, S.G., Sox, H.C., and Tversky, A.: 1982: "On the elicitation of preferences for alternative therapies," New England Journal of Medicine 306:1259-1262.

Merkhofer, M. and V. Covello: 1984, Risk Assessment and Risk Assessment Methods: The State-of-the-Art, Washington, D.C.: National Science Foundation.

Miller, J.D.: 1985, The Attitudes of Religious, Environmental, and Science Policy Leaders Toward Bioltechnology, Unpublished manuscript.

Morgan, M.G., P. Slovic, I. Nair, D. Geisler, D. MacGregor, B. Fischhoff, B. Lincoln, and K. Florig: 1985, "Powerline Frequency Electric and Magnetic Fields: A Pilot Study of Risk Perceptions," Risk Analysis, Vol. 5: 139-150.

National Academy of Sciences: 1987, Introduction of Recombinant DNA-engineered Organisms into the Environment: Key Issues. Committee on the Introduction of Genetically Engineered Organisms into the Environment. Washington, D.C.: National Academy Press.

National Research Council/National Academy of Sciences: 1983, Risk Analysis in the Federal Government: Managing the Process, National Academy Press, Washington, D.C.

National Safety Council: 1982. Accident Facts. Chicago: National Safety Council.

National Science Foundation: 1983, Science Indicators, (NSB 83-1), National Science Foundation, Washington, D.C.

National Science Foundation: 1985, Science Indicators--The 1§ Report, (NSB 85-1), National Science Foundation, Wash., D.C.

Nelkin, D.: 1984, Science in the Streets. New York: Twentieth Century Fund.

Nisbett, R.E., and L.D. Ross: 1980, Human Inference: Strategies and Shortcomings of Social Judgment, Englewood Cliffs: Prentice Hall.

Otway, H. J.: 1980 "Risk perception: A psychological perspective". In M. Dierkes, S. Edwards, and R, Coppock (eds.) Technological Risk: Its Perspective and Handling in Europe, Boston: Oelgeschlager, Gunn and Hain.

Otway, H., Maurer, D. and Thomas, K.: 1978, "Nuclear power: The question of public acceptance". Futures, 10, 109-118.

Otway, H.J. and v. Winterfeldt, D.: 1982 "Beyond acceptable risk : on the social acceptability of technologies". Policy Sciences, 8, 127-152.

N. Pfund and L. Hofstadter: 1981, "Biomedical Innovation and the Press," J. Commun. 31, 138-154 (1981).

Pochin, E.E.: 1975, "The Acceptance of Risk," British Medical Bulletin, Vol. 31(3):184-190.

President's Commission on the Accident at Three Mile Island (1979), Report of the Public's Right to Information Task Force. Washington, D.C.: U.S. Government Printing Office.

Renn, O. Man, technology, and risk: A Study on intuitive risk assessment and attitudes towards nuclear power, (Report Jul-Spez 115, Julich). Nuclear Research Center, 1981.

Rogers, E.M. and F.F. Shoemaker: 1971. Communication of Innovations: A Cross-Cultural Approach. New York: The Free Press.

Ruckelshaus, W.D.: 1983, "Science, Risk, and Public Policy," Science, 221: 1026-1028.

Ruckelshaus, W.D.: 1984, "Risk in a Free Society," Risk Analysis, September 1984, ppp. 157-163.

Ruckelshaus, W. D.: 1987, "Communicating About Risk," pp. 3-9 in J.C. Davies, V.T. Covello, and F.W. Allen (eds.). Risk Communication. Washington, D.C.: The Conservation Foundation, 1987.

Press, F.: 1987, "Science and Risk Communication," pp. 11-17 in J.C. Davies, V.T. Covello, and F.W. Allen (eds.). Risk Communication. Washington, D.C.: The Conservation Foundation, 1987.

Sandman, P.: 1986, Explaining Environmental Risk. Washington, D.C.: U.S. Environmental Protection Agency, Office of Toxic Substances.

Sandman, P.: 1986, "Getting to Maybe: Some Communications Aspects of Hazardous Waste Facility Siting," Seton Hall Legislative Journal, Spring 1986.

Sandman, P.M.: 1986, Explaining Environmental Risk. Washington, D.C.: U.S. Environmental Protection Agency, Office of Toxic Substances.

Sandman, P., D. Sachsman, M. Greenberg, M. Gochfeld: 1987, Environmental Risk and the Press. New Brunswick: Transaction Books.

Sandman, P., D. Sachsman, and M. Greenberg: 1987, Risk Communication for Environmental News Sources. Industry/University Cooperative Center for Research in Hazardous and Toxic Substances: New Brunswick, New Jersey.

Sharlin, H.: 1987, "EDB: A Case Study in the Communication of Health Risk," in B. Johnson and V. Covello, (eds.), The Social and Cultural Construction of Risk, Boston: Reidel.

Short, J.: 1984, "The social fabric of risk". American Sociological Review, December.

Slovic, P. Perception of risk. Science, 236, 280-285. 1987

Slovic, P., Fischhoff, B., and Lichtenstein, S. Accident probabilities and seat belt usage: a psychological perspective. Accident analysis and prevention, 10, 1978, 281-285.

Slovic, P., Fischhoff, B. and Lichtenstein, S.: 1980, "Facts and fears: Understanding perceived risk." In R. Schwing and W.A. Albers (Eds.), Social Risk Assessment: How Safe is Safe Enough? New York: Plenum, 1980.

Slovic, P., and B. Fischhoff: 1982, "How Safe is Safe Enough? Determinants of Perceived and Acceptable Risk," in Gould and Walker (eds.), Too Hot to Handle, Yale University Press, New Haven.

Slovic, P., B. Fischhoff, and S. Lichtenstein: 1980, "Informing People about Risk," in M. Maziz, L. Morris, and I. Barofsky (eds.), Banbury Report 6, The Banbury Center, old Spring Harbor, New York.

Slovic, P., B. Fischhoff, and S. Lichtenstein: 1982, "Facts versus Fears: Understanding Perceived Risks (Revision)," in Kahneman, D., P. Slovic, and A. Tversky (eds.): 1982, Judgment Under Uncertainty: Heuristics and Biases, New York: Cambridge University Press.

Slovic, P., Fischhoff, B., and Lichtenstein, S.: 1984, "Characterizing perceived risk." In R. W. Kates, C. Hohenemser & J. X. Kasperson (Eds.), Perilous progress: Technology as hazard. Boulder, CO: Westview, 1984.

Sjoberg, L.: 1979, "Strength of Belief and Risk," Policy Sciences 11, 39-57.

Smith, V.K., W.D. Desvousges, and A. Fisher: 1987, Communicating Radon Risk Effectively: A Mid-Course Evaluation., Report No. CR-811075. Washington, D.C.: U.S. Environmental Protection Agency, Office of Policy Analysis.

Starr, C.: 1969, "Social Benefit Versus Technological Risk," Science 165, 1232-1238.

Thomas, L. M.: 1987, "Why We Must Talk About Risk," pp. 19-25 in J.C. Davies, V.T. Covello, and F.W. Allen (eds.). Risk Communication. Washington, D.C.: The Conservation Foundation, 1987.

Tversky, A., and Kahneman, D.: 1981, "Judgment under uncertainty: Heuristics and biases." Science, 211, 1453-1458, 1981.

U.S. Congress, Office of Technology Assessment: 1987, New Developments in Biotechnology--Background Paper: Public Perceptions of Biotechnology, OTA-BP-BA-45 (Washington, DC: U.S. Government Printing Office, May 1987).

U.S. Environmental Protection Agency: 1984, Risk Assessment and Management: Framework for Decision Making, Report no. EPA/600/9-85/002. Washington, D.C.: U.S. Environmental Protection Agency. December, 1984.

U.S. Environmental Protection Agency: 1987, Unfinished Business: A Comparative Assessment of Environmental Problems. Vol. 1. Overview Report. Washington, D.C.: U.S. Environmental Protection Agency, Office of Policy Analysis, February, 1987.

Vlek, C. and Stallen, D. J. Judging risks and benefits in the small and in the large, Organizational Behavior and Human Performance, 28, 1981, 235-271.

Von Winterfeldt, D. and W. Edwards: 1984, "Patterns of Conflict About Risky Technologies," Risk Analysis, Vol. 4(1): 55-68.

Weinstein, N.D.: 1979, "Seeking reassuring or threatening information about environmental cancer". Journal of Behavioral Medicine, 2: 125-139.

Weinstein, N.D.: 1980, "Unrealistic Optimism about Future Life Event," Journal of Personality and Social Psychology 39(5) 806-820.

Weinstein, N. D.: 1984, "Why it won't happen to me: Perceptions of risk factors and susceptibility". Health Psychology, 3, 431-457.

Wilson, R.: 1979, "Analyzing the Daily Risks of Life," Technology Review 81, 40-46.

Wilson, R.: 1984, "Commentary: Risks and Their Acceptability," Science, Technology, and Human Values 9 (2), Spring, 11-22.

Wilson, R. and E. Crouch: 1987, "Risk Assessment and Comparisons: An Introduction." Science, Vol. 236 (17 April 1987): 267-270.

Young, F. and H. Miller: 1987, "The NAS Report on Deliberate Release: Toppling the Tower of Bio-babble," Bio/Technology 5 (October 1987): 1010.

PART III

SCIENTIFIC PERSPECTIVES

APPLICATION OF GENETICALLY - ENGINEERED MICRO-ORGANISMS IN THE ENVIRONMENT

Ben Lugtenberg, Letty de Weger and Carel Wijffelman
Leiden University, Department of Plant Molecular Biology,
Nonnensteeg 3, 2311 VJ Leiden, The Netherlands

INTRODUCTION

The application of micro-organisms in the environment for beneficial purposes, e.g. as microbial pesticides or as natural fertilizers, has a long history. The Enviromental Protection Agency registered the microbial pesticides Bacillus popillae (to control Japanese beetle larvae) in 1948, B.thuringiensis (to control moth larvae) in 1961 and Agrobacterium radiobacter (to control crown gall caused by A.tumefaciens) in 1979. Rhizobium bacteria have been available as seed inoculants in the USA since 1896.

The use of micro-organisms for such purposes has recently attracted much attention now recombinant DNA techniques are used for the improvement of currently used strains as well as for new applications. It is clear that the use of more powerful and new strains could be very beneficial for mankind. On the other hand the major concern of some scientists is that genetic modifications might convert non-pathogens to pathogens, alter the host range or virulence or that new species are being created by recombinant DNA technology which could cause environmental disasters. The current idea in the United States and in most European countries is that application of genetically-engineered micro-organisms will only be allowed after a careful risk evaluation has shown that tests carried out to detect possible adverse effects yield negative results. The decision to allow a certain application should be based on a careful weighing of on

NATO ASI Series, Vol. G18
Safety Assurance for Environmental Introductions
of Genetically-Engineered Organisms
Edited by J. Fiksel and V. T. Covello

one hand the expected advantages for the people and for the company and on the other hand the estimated risks for a smaller or larger part of mankind. If decision-makers are convinced that the benefits outweigh the risks, application should be allowed as fast as possible. Unnecessary requirements for additional experiments are frustrating, extremely expensive and might result in quitting the development of projects on microbial pesticides.

BENEFITS OF THE APPLICATION OF MICRO-ORGANISMS IN THE ENVIRONMENT

Some of the major beneficial purposes of the application of micro-organisms in the environment are (i) to increase crop yield, thereby decreasing the world food problem, (ii) to avoid the adverse effects of <u>chemical</u> crop protection, (iii) to save artificial fertilizer thereby decreasing costs as well as the acid rain problem, (iv) to decrease frost damage, (v) to clean polluted water and soil, (vi) to reduce sulphur in coal, (vii) to recover metal (e.g. Cu) from ore. These and other examples of the application of micro-organisms in the environment will be discussed in the following paragraphs.

One third of the yield of crop plants is lost as a result of pests. Decrease of these losses will, especially in third world countries, contribute to abolishing the world food problem. Present applications of microbial pesticides are the fungus <u>Trichoderma</u> and several bacteria like <u>B.popillae</u> and <u>B.thuringiensis</u>, which control larvae, and <u>A.radiodurans</u>, which protects several plants against the crown gall- causing bacterial pathogen <u>A.tumefaciens</u>.

<u>1. Bacillus thuringiensis</u>

<u>B.thuringiensis</u> produces a toxin, designated as δ-endotoxin or

B.t. toxin, by which it is able to control lepidopteran insect pests. This biocontrol agent has the disadvantage that the applied bacteria die too soon. Therefore the gene coding for the toxin has been cloned and expressed in a plant root-colonizing Pseudomonas fluorescens isolate which is thought to be a better delivery system for the toxin. Although growth chamber experiments are highly encouraging a true test of the effectiveness of this biocontrol agent requires field tests, for which permission has not yet been granted (Graham et al 1986).

2. Ice-nucleating bacteria

Some bacteria of the genera Erwinia, Pseudomonas and Xanthomonas can nucleate the crystallization of ice from super-cooled water. P.syringae produces a protein which acts as a crystallization center for ice (Wolber et al 1986). The resulting ice crystals are responsible for considerable frost damage to different crop plants. By deleting the gene coding for this ice-nucleation protein a mutant was constructed in the hope that, when sprayed on the plants, it would be able to prevent colonization of the plant tissue with frost damage-causing wild type strains. The first field trials with this "ice-minus" bacterium were carried out in 1987. Although, in comparison with the naturally-occurring wild type varieties, the mutant strain does not contain additional DNA but instead has lost part of its DNA, there has been considerable opposition against small scale field trials. However, from the point of view of a molecular microbiologist it is extremely unlikely that this deletion mutant has obtained properties which could be harmful for the environment.

3. Agrobacterium radiodurans strain 84

Crown gall is a plant disease caused by the soil bacterium A.tumefaciens which enters the plant (e.g. almond, peach, rose, etc.) through wounds. Since 1973, commercial stone fruit and rose growers in Australia have protected their crops by dipping their planting material in a suspension of a related bacterium,

A.radiodurans, thereby achieving nearly complete control of the disease (Kerr 1980). The particular A.radiodurans strain 84 produces a toxin, designated as agrocin 84, which kills the pathogenic bacterium. The genetic code for agrocin 84 is located on a plasmid. Also in A.tumefaciens the pathogenic properties as well as its sensitivity to agrocin 84 are located on a plasmid, the Ti (tumor-inducing) plasmid. Under laboratory conditions derivatives of the pathogen which are resistent to the toxin can easily be obtained, provided that nopaline is present. Such derivatives are no longer subject to biological control by strain 84. In the first seven years of application such cases have not been detected in the field in Australia. However, in Greece A.tumefaciens strains resistant to agrocin 84 have been isolated from peach (Kerr and Panagopoulos 1977). Moreover, many of these resistant A.tumefaciens isolates even produced agrocin 84. Appar'ently genes controlling agrocin 84 production have been transferred from strain 84 to a pathogenic recipient (Panagopoulos et al 1979). The potential advantages of genetic engineering to prevent these problems are clear. Firstly, the transfer gene(s) can be deleted from the beneficial strain thus preventing transfer, and secondly, the gene(s) controlling agrocin 84 production can be placed on a plasmid which is incompatible with the Ti plasmid, thus preventing the stable maintainance of the two traits agrocin 84 production and pathogenicity. Combination of the transfer-negative property with incompatibility would almost certainly prevent the occurrence of agrocin 84-producing pathogenic strains.

4. Trichoderma hamatum

Chemical protection of crop plants is currently necessary in modern crop management. Chemical control is cheap but adverse effects of such chemicals are also evident. Therefore scientists have been looking for biological alternatives as control agents.

A group of diseases, collectively known as damping-off, is common in both commercial nurseries and in amateur gardening. Several common soil fungi are involved, especially Pythium, Fusarium and Rhizoctonia solani. These attack weak seedlings or germinating seeds, especially in the cold conditions in early spring which reduce plant growth. These diseases can be controlled by the use of fungicidal seed dressings or by sterilizing the soil in which the seeds are sown. Recently, biological control agents are available commercially. Usually the fungus Trichoderma hamatum is inoculated into the soil and kills or antagonizes the pathogens (Campbell 1985). The same pathogens can be attached by fluorescent Pseudomonas bacteria (see later on). Practical biological control with the fungus Trichoderma hamatum has been plagued by erratic results. In at least one case the failure to protect pea seeds against Pythium spp. under low Fe^{3+} conditions is caused by antagonism by fluorescent soilborne seed-colonizing pseudomonads (Hubbard et al 1983).

5. Fluorenscent Pseudomonas spp

In the last decade considerable progress has been made with the application of fluorescent Pseudomonas bacteria under low iron conditions (Kloepper et al 1980, Schroth and Hancock 1983). Plant growth stimulation has been observed in greenhouses, in aquaculture and in soil.

Fluorescent root-colonizing Pseudomonas bacteria are antagonistic to a wide range of plant-pathogenic bacteria and fungi (Kloepper and Schroth 1981, Geels and Schippers 1983a, O'Gara et al 1986, Scher 1986).

Take-all and Pythium root rot, incited respectively by Gaeumannomyces graminis and several Pythium species are major constraints to wheat production. Seed treatment with selected fluorescent pseudomonads has been tested in experimental plots in the United States and in South Australia against the take-all fungus. The bacterial treatment appeared to suppress take-all and increased yield from 5-21% (Weller and Cook 1986).

Fusarium wilt incidence of flax, radish, cucumber and carnation can be reduced by P.putida (Scher 1986).

Detailed field experiments carried out in The Netherlands have shown a considerable decrease in potato tuber yield with increasing cropping frequency leading to yield decreases of

15-30%. Physical and chemical soil factors and regular soil-borne pathogens do not fully account for the observed reductions in yield. Currently the attention focusses on HCN-producing pseudomonads as a major agent responsible for yield reduction (Schippers et al 1986). HCN is known to inhibit cytochrome oxidase respiration and thereby the energy metabolism of plant cells (Lambers 1980). The observed yield reductions can be reproduced in pot experiments in soil taken from the relevant cropping frequency field plots. If potato seed tubers were dipped in cell suspensions of selected fluorescent isolates P.fluorescens WCS374 or P.putida WCS358 before sowing, these tuber yield reductions could be prevented in pot experiments (Geels and Schippers 1983b). Field experiments carried out so far had varying degrees of success (Schippers et al 1986).

The fluorescent properties of the beneficial Pseudomonas strains are caused by the production of siderophores, Fe^{3+}-binding molecules, which are secreted by several micro-organisms under conditions of low Fe^{3+} (Neilands 1982, Meyer and Abdallah 1978, Marugg et al 1985). Secreted siderophores scavenge traces of Fe^{3+} which is required for growth. The resulting Fe^{3+} siderophore complex is taken up by the micro-organism by means of highly specific protein receptor molecules located at the bacterial cell surface (Neilands 1982). The beneficial Pseudomonas strains produce particularly effective siderophores which are thought to limit the availability of Fe^{3+} for pathogens and other harmful micro-organisms, but not for the plant (de Weger et al 1987a). Becker et al (1986) have elegantly shown that the bacterial siderophore agrobactin is able to stimulate Fe^{3+} uptake by plants. The result is that pathogenic and other harmful microbes decrease in number which is beneficial for plant growth (de Weger et al 1987a).

In cases that beneficial effects were not observed, colonization of the root surface by Pseudomonas had usually been weak, suggesting that the "siderophore delivery system" had not been functioning effectively. By improving this delivery system antagonistic pseudomonads might become effective microbial pesticides. Until recently hardly anything was known about factors responsible for root colonization. Several lines of evidence indicate that proteolytic activity (O'Gara et al 1986) and motility (de Weger et al 1987b) of fluorescent Pseudomonads are important factors for root colonization.

Scientists are presently aiming at improving the siderophore production and the root colonization properties of the beneficial pseudomonads by molecular genetic techniques in order to transfer the laboratory scale successes to reproducible yield increases under field conditions.

6. Rhizobium

One of the best examples both in terms of beneficial effect as well as in terms of period of experience is the use of Rhizobium bacteria. These can invade leguminous plants and form nodules at the plant's roots. The Rhizobium present in these nodules is able to convert atmospheric nitrogen into ammonia which is used by the plant as a major nitrogen source. The application of Rhizobium decreases the need for artificial fertilizer, which is produced by a process which demands a lot of energy. Some 16 million metric tons of nitrogen fertilizer are applied annually world-wide, which is projected to increase to 160 million metric tons by the year 2000 (OECD 1986). Denitrification, i.e. conversion of nitrate to N_2 or N_2O, occurs after application of nitrogen fertilizer. Depending on the type of soil up to 37% of the applied nitrogen can be lost by denitrification (Campbell 1985). This fraction is not only lost for beneficial purposes, it also contributes substantially to the acid rain problem.

Rhizobium bacteria and/or leguminous plants are fastidious in that a certain bacterium can fix nitrogen on only a limited number of the leguminous plants, i.e. they have a very limited host range.

Rhizobium is presently studied by a large number of research groups. The growing interest is stimulated by (i) the importance of biological nitrogen fixation for agriculture and (ii) the hope that by genetic engineering the amount of fixed nitrogen per plant can be increased or even that the ability to fix nitrogen can be transferred to non-leguminous plants. The results obtained so far have shown that nodulation and nitrogen fixation are extremely complex processes (Long 1986). Increase of nitrogen fixation looks feasible but no plant-bacterium combinations have yet been constructed that give higher fixation potential than is normally found (Hobbs et al 1986). However, the nitrogen fixation process is too complex to expect that the nitrogen-fixing ability can soon be transferred to other plants.

Inoculation of seeds with Rhizobium bacteria is a common practice in the United States for seeds of soybean and Alfalfa. One of the problems is that these introduced Rhizobium strains, which were chosen mainly on the basis of their excellent capacity to fix nitrogen, compete poorly with indigenous strains (Stacey and Brill 1982). Moreover, studies on field-grown soybeans indicate that the most competitive inoculum strains in one soil may be the least effective in another (Ham et al 1976).

Presently studies directed to improve nitrogen fixation focuss on (i) the construction of strains that fix more nitrogen, (ii) understanding the process of competition between inoculum strains and indigenous rhizobia in order to solve this problem and (iii) to understand the phenomenon of host specificity in order to be able to construct Rhizobium strains that can be used for a wider spectrum of plants.

7. Azospirillum

Free-living N_2-fixing bacteria of the genus Azospirillum are found associated with the roots of many tropical grasses and cereal crops. They have beneficial effects on crop yields and decrease fertilization requirements (Neyra and Dobereiner 1977). The observed correlation between enhanced ion uptake and inoculation of corn with Azospirillum brasilense has led to the suggestion that ion uptake may be a significant factor in the crop yield enhancement (Lin et al 1983).

8. Agrobacterium rhizogenes

A.rhizogenes is a soil bacterium which can enter roots via wounds or natural openings which may result in a proliferation of secondary roots ("hairy root syndrome"). Infectivity is correlated with a large Ri (root inducing) plasmid. A possible application of A.rhizogenes is in drought tolerance (Moore et al 1979). Therefore the possibility of using A.rhizogenes in agriculture because of its root-initiating activity was tested. Application to bare root stock almond trees indeed showed a positive effect on root number and root mass without pathological effects (Strobel and Nachmias 1985).

9. Biodegradation of toxic chemicals

Biodegradation is the use of micro-organisms or their metabolites to treat toxic chemicals and render them harmless. Biodegradation is potentially less expensive than any other approach to neutralizing toxic wastes. As early as 1914, researchers were isolating strains that efficiently degrade noxious sewage and reduce the volume of organic matter. Today, micro-organisms that degrade some of the most toxic of chemicals are being isolated from the soil of hazardous-waste sites (Nicholas 1987).

One of the best known released toxic chemicals is 2,4,5-trichlorophenoxyacetic acid (2,4,5-T), which has been used extensively as a component of Agent Orange in Vietnam and as a herbicide in various countries. The compound has created toxicological problems (Grant 1979). A possible solution for the very slow rate of biodegradation is the isolation of a strain of Pseudomonas cepacia which could degrade as much as 95% of the 2,4,5-T in soil within one week (Chatterjee et al 1982).

10. Extraction of metals and sulphide

Micro-organisms have been commercially used in the recovery of metals from ore for over a century. In the 1950s it was shown that Thiobacillus ferroxidans and T.thiooxidans oxidize some metals as well as several copper sulphide minerals. These micro-organisms are used today to leach copper and uranium from ores in significant commercial quantities (OECD, 1986) and to decrease the sulphur content of coal.

11. Oil removal and recovery

A considerable amount of the world's oil reserves remain in subterranean wells where the oil is either trapped in the rock formation or is too viscous to pump. Researchers are developing new processes, including microbiological ones, to recover more of the oil. The microbial polymer emulsan (Pines and Gutnick 1981) produced by Acinetobacter calcoaceticus can be used to enhance removal of residual oil from tanker holds. Emulsan is expected to undergo testing as an oil recovery enhancer which would help make the oil easier to pump from a well (OECD 1986).

CONSIDERATIONS ON THE APPLICATION OF GENETICALLY-ENGINEERED MICRO-ORGANISMS IN THE ENVIRONMENT

1. Application of micro-organisms in the environment is not new. In contrast, the large number of examples we have experience with should give us confidence in future applications with genetically-engineered micro-organisms. Many of the present applications have proven to be beneficial to mankind, and not one case of accident has been reported. The worse thing that has happened is that in Greece A.tumefaciens has become immune to treatment with A.radiodurans but the resulting situation is hardly worse that the situation before 1973 when the crown gall disease caused by A.tumefaciens could not be treated at all.
2. Genetically-engineered micro-organisms are not new. Scientists in hundreds of laboratories all over the world have genetically manipulated microbes for more than a decade. The benefits are clear now several important products have been marketed already whereas others are in the pipeline. Despite predictions about major risks to mankind and to the environment not one report on an accident has appeared.
3. New is the combination of 1 and 2, the application of genetically-engineered micro-organisms in the environment. Since experience with this combination is lacking, good scientific

evidence for the possibility of adverse effects does not exist. Therefore it is reasonable to consider this as a new situation on which the normal process of risk analysis, consisting of the steps (1) risk identification, (2) risk-source characterization, (3) exposure assessment, (4) dose-response assessment, and (5) risk estimation (Fiksel and Covello 1986) should be applied.

4. In the process of risk analysis one should take into consideration that the various potential applications differ from each other with respect to experience with related cases, possible adverse effects and possible benefits. Care should be taken to define the relevant questions correctly and to design those experiments that are likely to give the best answer. Questions should focuss on those scientifically established facts that are as closely as possible related to the properties of the released micro-organism and its DNA in the relevant environment. Questions should not be diverted from the subject matter by comparing this issue with unrelated accidents (e.g. Bhopal, Tsjernobyl). It is important to reach consensus on the important questions among the scientists (see 5) and to involve representatives of the local community in an early stage. Even in a case-by-case approach it is likely that the procedure for the application of genetically-engineered micro-organisms in soil will at least include studies on (i) the persistence and spread of the micro-organism, (ii) colonization of relevant crop plants, (iii) influence of the presence of crop plants on bacterial numbers (iv) transfer of manipulated DNA to indigenous pathogens and to bacteria related to the applied one.

5. In a good risk assessment process, the advice of experts from the various disciplines involved is essential. It is our major concern that the complexity of the subject is underestimated and that the studies are for example supervised only by ecologists or only by geneticists. In the case of application of a micro-organism the most important disciplines are likely to include molecular microbiology, microbial ecology, microbial and molecular genetics, plant pathology and soil biology. These may have to be extended with agronomy, animal and plant ecology, and epidemiology.

MONITORING THE FATE OF A GENETICALLY-ENGINEERED BACTERIUM AND OF DNA INTRODUCED INTO IT

In assessing possible risks related to the release of genetically-engineered micro-organisms it is of primary importance to monitor separately the fates of the introduced bacterium and of the genetically-manipulated DNA. The reason is that the bacterium and the introduced DNA are not necessarily unseparable entities since the introduced DNA can be lost by its host cell, be transferred to other microbes or be changed. Methods which can be used to recognize and to recover a bacterium and its DNA will be discussed below. Recognition of the bacterium and its DNA is greatly facilitated when, prior to release, appropriate tags are introduced.

ADVANTAGES AND DISADVANTAGES OF VARIOUS TAGS

Molecular tags can be of great help for the detection, identification, recovery and enumeration of micro-organisms or of a particular piece of DNA. Advantages and disadvantages of a number of possible tags are summarized in Table 1.

The use of tags in microbial ecology has a long history in that the use of antibiotic resistant derivatives for monitoring the behaviour of a particular strain has been a common practice. Since by definition antibiotics interfere with crucial metabolic routes, mutations in genes coding for their target molecules are likely to have a negative effect on the gene product's primary function. Indeed, it has been reported that resistance to rifampicin, which coincides with an altered RNA polymerase, is associated with a significant loss of nodulation competitiveness in Rhizobium meliloti. The authors suggest that rifampicin resistance is unsuitable as a marker for ecological studies (Lewis et al 1987). The results of another study (Geels and Schippers 1983a) strongly suggest that derivatives of fluorescent

TABLE 1. EVALUATION OF POSSIBLE TAGS

TAG	RECOVERY BY PLATE SELECTION	SCIENTIFIC VALUE FOR MONITORING	POSSIBLE RISK
Classical antibiotic resistance	Yes	Limited[a]	None
Transposon	Yes	High[a]	Mutagenesis[a]
Non-mutagenic, non-selectable, not naturally occurring tags	No	Limited[a]	Very low

[a] See text for comments

pseudomonads, which were made resistant to the antibiotics rifampicin and nalidixic acid, have a selective disadvantage.

Since transposons can jump to random positions in the DNA, it is conceivable that among a collection of transposon mutants strains exist with mutations in unimportant or silent genes which therefore do not affect the strain's behaviour. From the scientific point of view such tags are ideal because they are very stable, have no selective disadvantage and they can easily be detected since antibiotic resistance can easily be selected for. This selection can also be used in case the transposon has been transferred to another bacterium, provided that it expresses the antibiotic resistance. Since a transposon is also a mutagen, some people feel that it should not be released. We share this objection only for large scale applications and we are tempted to believe that the use of transposons for monitoring purposes will not significantly contribute to the mutagenic capacity that is already present in the environment, e.g. ultraviolet light and naturally occurring mutagens, among which transposons. In principle it is also possible to delete the transposase activity of a transposon, thereby deleting its mutagenic character.

It has been argued that, to circumvent the mutagenic character of transposons, other tags should be used. DNA coding for a protein product that does not occur in the environment could be such a tag. Alternatively, DNA rich in nonsense codons in all reading frames would even be safer since it would prevent its own expression. Although such approaches can result in super-safe strains, they have the tremendous disadvantage that strains carrying these tags cannot be selected on antibiotic plates or otherwise. The low operational feasibility of these tags is a clear weakness. An exception seems the introduction of the trait of lactose utilization into the chromosome of fluorescent pseudomonads. This manipulation had no detectable effects on the growth, survival and hardiness of the pseudomonads (Drahos et al 1986, Barry 1986).

IDENTIFICATION AND RECOVERY OF RELEASED BACTERIA

Identification based on the classical surface antigens

The classical serological identification of bacteria is mainly based on identification of K-, H-, and/or O-antigens. The structures or molecules corresponding with these antigens are listed in Table 2. Their position on the bacterial cell is schematically indicated in Fig. 1. Agglutination techniques are often used to detect these antigens. A drawback of this way of identification is that relatively many strains are "untypeable". This is not surprizing since capsule, fimbriae, flagella and O-antigens are not always present on bacterial cells.

TABLE 2. MAJOR BACTERIAL ANTIGENS

ANTIGEN DESIGNATION	CORRESPONDING STRUCTURE/MOLECULE
K	Capsule, fimbriae ("pili")
H	Flagella
O	Part of lipopolysaccharide

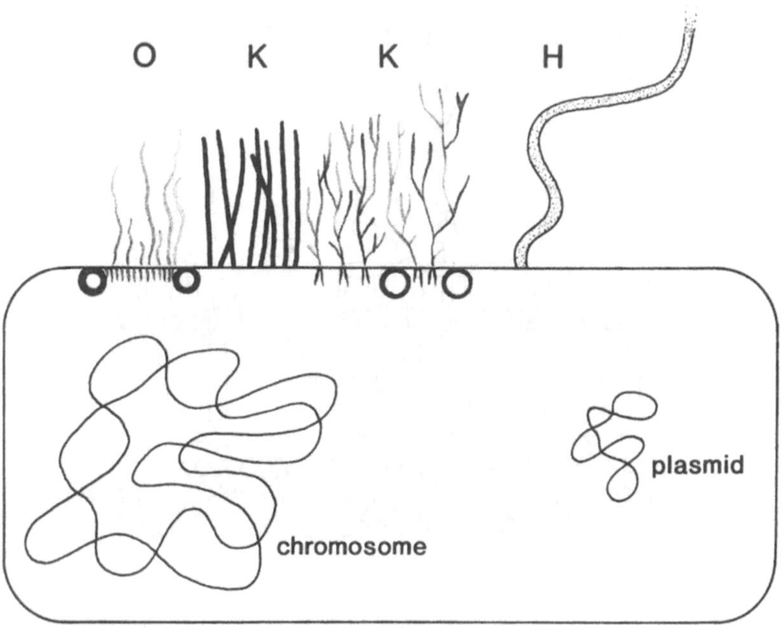

Fig. 1. Schematic representation of the major antigens of a Gram-negative cell. The O-antigen part of the lipopolysaccharide (LPS), the fimbrial antigen (left K), the capsular antigen (right K) and the H-antigen are all surface-exposed. The rings at the cell surface represent outer membrane proteins which can be shielded from immunoglobulins in whole cells when the O-antigen is present (van der Ley et al 1986). Chromosomal as the DNA is always present; plasmid DNA can be present.

An application of this serological identification is the use of fluorescent antibodies to follow the fate of a specific strain in the environment. In the case of a Gram-negative bacterium, the most specific antibodies are most likely directed against the highly immunogenic LPS molecule. A drawback of the use of fluorescent antibodies is that the technique cannot be used for enumeration since many bacteria are lost during the labelling procedure

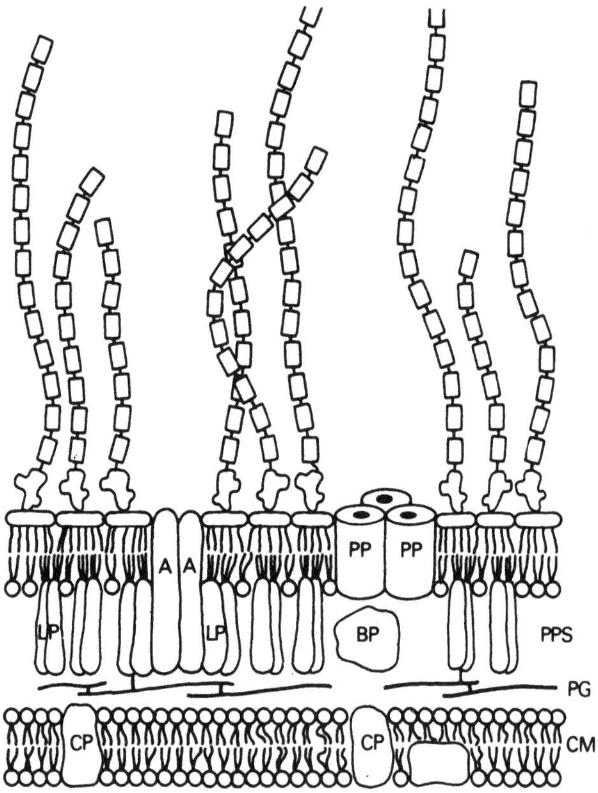

Fig. 2. Molecular architecture of the cell envelope of Enterobacteriaceae. The cytoplasmic membrane (CM) consists of a phospholipid bilayer in which proteins, among which carrier proteins (CP), are present. The outer membrane consists of an asymmetric lipid bilayer with the acyl chains of phospholipid and lipoprotein forming the inner leaflet. The outer lipid leaflet is predominantly formed by the lipid part of LPS. Three types of outer membrane proteins, namely pore protein (PP), OmpA protein (A) and lipoprotein (LP) are indicated. The two membranes are separated by a peptidoglycan layer (PG) and a periplasmic space (PPS) which contains binding proteins (BP). (After Lugtenberg and van Alphen 1983).

Identification based on electrophoretic patterns of cell surface components

Fig. 2. represents the present view on the molecular architecture of an important class of Gram-negative bacteria, the Enterobacteriaceae (Lugtenberg and van Alphen 1983). Two important features that we want to stress here are valid for all Gram-negative bacteria, although the details of the structure of the cell surface presented in Fig. 2 does not apply for all Gram-negative bacteria (Lugtenberg 1985).

(i) Upon sodium dodecyl sulphate polyacrylamide gel electrophoresis (SDS-PAGE), outer membrane proteins form a considerable number of predominant bands whose position in the gel is a reflection of their molecular weight (Fig. 3). Since the resulting outer membrane protein patterns carry a vast amount of information on the size and regulation of these proteins, it is not surprising that these patterns can differ between species and even between strains within a given species (Schnaitman 1974; Lugtenberg et al 1976; Overbeeke and Lugtenberg 1980; Achtman et al 1983; Lugtenberg et al 1984; Lugtenberg et al 1986; Dijkshoorn et al 1987).

The fact that information about outer membrane proteins has so far not often been used to discriminate between related strains is most likely due to the fact that many outer membrane proteins are not accessible to antibody molecules in whole cells since they are shielded by the O-antigen of LPS (see van der Ley et al 1986).

(ii) LPS (lipopolysaccharide) is usually composed of three parts, designated as lipid A, the core and the O-antigen (Fig. 4). Upon SDS gel electrophoresis LPS of a pure culture of a given strain often appears to be extremely heterogeneous since it splits into a large number of bands (Fig. 5), most of which correspond with molecules with an increasing number of O-antigen repeating units (Goldman and Leive 1980; Palva and Mäkelä 1980; Hitchcock and Brown 1983).

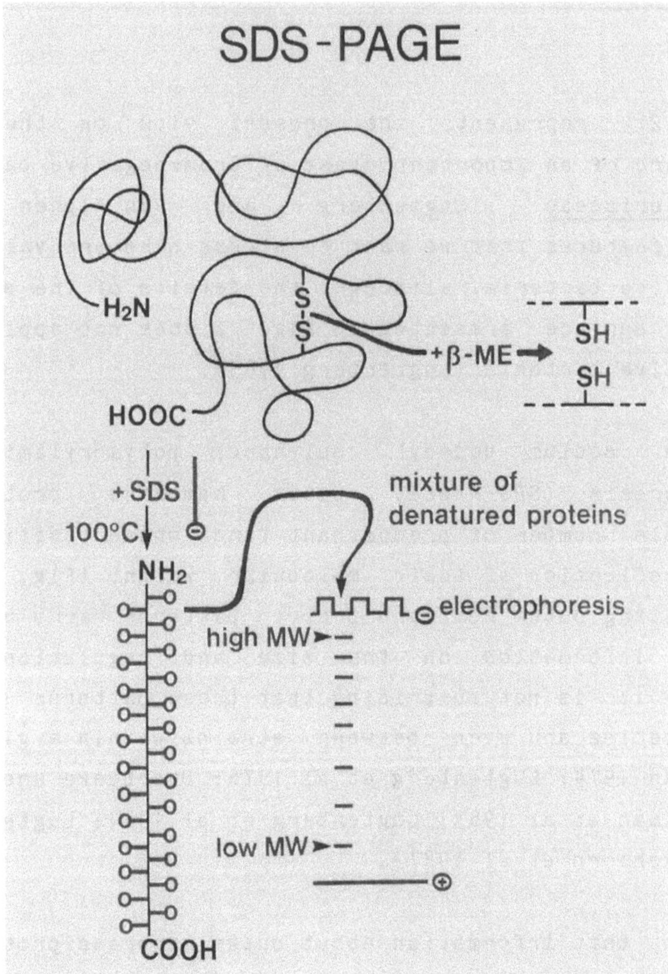

Fig. 3. Background of sodium dodecylsulphate polyacrylamide gel electrophoresis. Protein molecules, which may contain -S-S-bridges, are treated (i) with the reducing agent β-mercapto-ethanol (β-ME), (ii) with SDS and (iii) heated. β-ME opens the -S-S- bridge thereby allowing so the protein to unfold completely. The hydrophobic acyl chain of SDS binds to the hydrophobic amino acid residues of the protein up to a weight/weight ratio of SDS to protein of 1.4:1.0. All protein molecules are thereby converted to rods. The negatively charged SDS molecules give all proteins a negative charge. Additional heating is required to accomplish complete unfolding of some proteins. The complete treatment mentioned so far is simply carried out by mixing a protein sample with a sample mixture (containing a.o. SDS and β-ME) followed by a brief incubation in a water bath. The resulting solution is subsequently applied on top of a polyacrylamide gel. After electrophoresis, which separates the proteins according to their sizes, the proteins are visualized by staining, usually with Coomassie Brilliant Blue, Fast Green or with a silver-containing reagent.

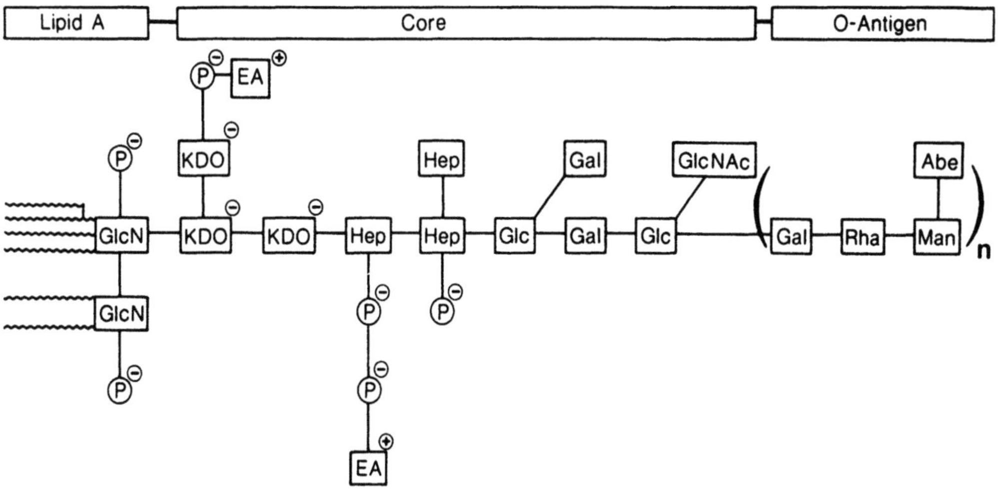

Fig. 4. Structure of lipopolysaccharide. The three parts, lipid A, core and O-antigen, are indicated. The lower figure is a schematic representation of the chemical structure of LPS from Salmonella typhimurium. Wavy lines represent long chain fatty acids (C8 - C16). The number of repeating O-antigen units, is variable, e.g. zero in Escherichia coli K_{12} and up to 30 to 100 in many wild strains. Note that many residues in the lipid A-core region of the molecule are charged. The phosphate residues on the glucosamines are sometimes substituted. Abbreviations: Abe, abequose; EA, ethanolamine; Gal, galactose; Glc, glucose; GlcN, glucosamine; GlcNAc, N-acetyl glucosamine; Hep, L-glycero-D-manno-heptose; KDO, 2-keto-3-deoxy- octulosonic acid; Man, mannose; P, phosphate; Rha, rhamnose.

The heterogeneity of LPS within one culture (and therefore probably within a single cell) results in a ladder pattern of these various LPS species on a SDS-polyacrylamide gel. Differences in the O-antigen part of LPS have been used extensively for epidemiological studies on the spread of a particular strain. Therefore it is not surprising that LPS ladder patterns observed on gels differ between strains (Fig. 6) (e.g. Van der Ley et al 1986).

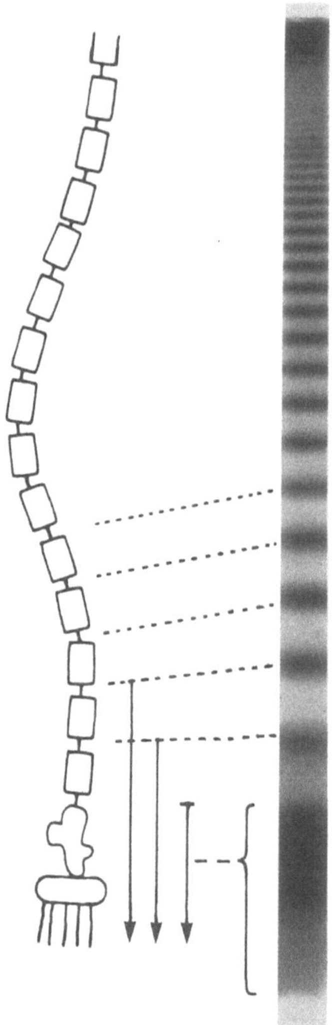

Fig. 5. Correlation between LPS structure and ladder pattern observed after SDS-PAGE. A schematic representation of the LPS structure (see also Figs. 2 and 4) is tentatively correlated with the ladder pattern observed upon SDS-PAGE of LPS of a fluorescent Pseudomonas strain. The lowest, fastest running, band most likely correlates with the complete lipid A-core region without O-antigen. The second step of the ladder represents the same structure substituted with one O-antigen repeating. Each higher step on the ladder is supposed to contain one additional O-antigen repeating unit.

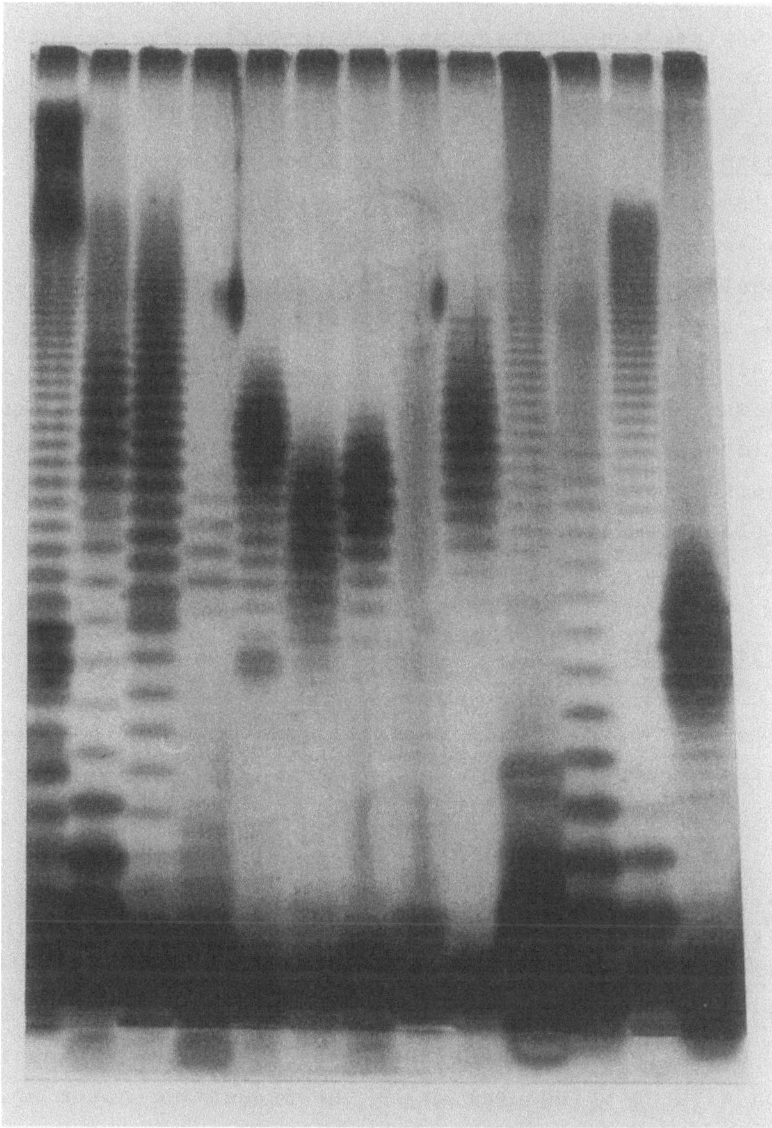

Fig. 6. Silver-stained patterns of proteinase K-treated membranes obtained after SDS-PAGE. The proteinase K treatment degrades the membrane proteins. Most bands observed upon subsequent staining represent LPS. Lanes represent, from left to right, samples of fluorescent Pseudomonas WCS strains 375, 007, 085, 134, 307, 314, 315, 324, 326, 361, 365, 366 and 379.

The techniques for characterizing strains on the basis of their outer membrane protein and LPS patterns have recently been applied to soil bacteria. Using a collection of 24 independent isolates of fluorescent, antagonistic, root-colonizing *Pseudomonas* isolates, it could convincingly be shown that all 24 tested isolates are different from each other (de Weger et al 1986; de Weger et al 1987c).

The analysis of outer membrane protein and LPS profiles provides a powerful addition to the techniques for strain identification. Several hundreds of isolates can be screened by one person in one week. The sensitivity of the technique can even be improved further when it is combined with the sensitivity and specificity of antibodies and it is then designated as Western blotting (e.g. Lugtenberg et al 1986).

Identification based on intracellular molecules

Profiles of total cell protein have been used instead of profiles from outer membrane proteins. Since the cell surface is subject to variation due to selective pressure of phages, bacteriocins etc., we feel that the profile of surface-exposed proteins, which constitute 10-20% of the total cell protein, is more likely to yield differences between strains than that of total cell protein.

Isozyme profiles can also be used to discriminate between strains. The proteins are either separated on a non-denaturing gel or on a denaturing gel with subsequent renaturing. The number of visible bands, and therefore also the potential to discriminate between strains, is usually low.

Strains can be characterized on the basis of their plasmid profiles. The number of plasmids, and therefore the chance to detect differences, is usually low. Some strains have no detectable plasmids at all.

The best method to discriminate strains on the basis of intracellular molecules is the use of DNA restriction fragment patterns. Restriction fragments are separated according to their size and specific fragments can be detected using radio-labelled probes which hybridize with certain complementary nucleotide sequences.

Recovery of introduced bacteria

Bacteria which do not carry a selectable tag can be recovered from environmental samples by the use of specific antibodies. The most difficult form of recovery is to attempt the isolation of bacteria in the physiological growth stage of the environment, since this requirement does not allow growth of the cells during the isolation procedure. If necessary, e.g. in soil samples, the bacteria should first be brought into suspension quantitatively. Subsequent concentration and/or selection could be brought about by "fishing" with monospecific antibodies covalently linked to a carrier e.g. polystyrene. Alternatively, the concentrated cells could first be labelled using fluorescent antibodies and, using a fluorescence-activated cell sorter, the positive cells could be separated from the negative ones. Analysis of the surface proteins of bacteria which still have the chemical composition that they had in the environment, could give a useful indication of the growth conditions (e.g. nutrient limitation) in that environment (Lugtenberg 1985). The described approaches for the isolation of the bacteria will only work if the antigen (i) is produced under the tested environmental conditions and (ii) is surface-exposed under these conditions.

If selectable tags are present, special plates can be used to select the introduced tag. It is important to realize that, depending on whether the bacterium or the introduced DNA is selected, either the fate of the cell or that of the DNA introduced into it, is followed.

RECOGNITION AND IDENTIFICATION OF INTRODUCED DNA

Possible fates of introduced DNA

Genetically-manipulated DNA in a released bacterium can be present either in the chromosome or in a plasmid (Fig. 1). The fate of the introduced DNA not only depends on its own stability but also on the fate of the bacterium and on whether it is actively transferred to other bacteria. The following events can take place.

1. The DNA in the "released" bacterium is stably maintained and is not transferred to other organisms, i.e. the introduced DNA remains intact and in the same position in the host DNA.
2. The introduced DNA is unstable without being transferred to other organisms. This can have the following reasons:

(a) The introduced DNA may undergo internal rearrangements, e.g. deletions or insertions.

b) The introduced DNA may undergo transposition, i.e. it is being transferred to another site in the host DNA.

(c) The introduced DNA is preferentially lost.

The presence of the introduced DNA can be detected by DNA-DNA hybridization. Using this technique it can also be checked whether this DNA itself undergoes changes or whether it is transferred to other sites in the host DNA, since all these changes will result in changed sizes of the appropriate restriction fragments.

3. The introduced DNA is transferred to another micro-organism.

Under laboratory conditions transfer of DNA to related micro-organisms can often be accomplished, provided that the appropriate conditions are offered. Specific plasmid "vectors" have been constructed to enable DNA transfer between various bacterial species. Unstable maintenance of the "foreign" DNA is then often a serious problem.

DNA transfer under more natural conditions is less well documented. The occurrence of conjugation between <u>Enterobacteriaceae</u> in the intestine of mice (Schneider et al

1961; Jones and Curtiss 1969) was clearly shown. Similarly, conjugation of E.coli in sterile soil and sterile stream water was detected (Weinberg and Stotzky 1972; Trevors and Oddie 1986). Gene exchange by transformation in sterile soil has been reported (Graham and Istock 1978, 1979). However, the value of the latter data for the natural situation has been questioned since unrealistically high bacterial cell concentrations were used (Sayler and Stacey 1986). Significant gene transfer in aquatic environments by transduction has been claimed to occur in P.aeruginosa (Miller et al 1986).

The studies mentioned are not truly representative for the natural environment where temperature, nutrient availability and pH are all dynamic parameters which can exert control over gene transfer. Also the presence of other competing (micro-)organisms will influence the behaviour of the studied bacteria. It is clear that as part of the risk assessment process the possible transfer of introduced DNA to other micro-organisms should be considered. It is equally clear that the methodology to study gene transfer in natural systems, especially in soil, needs considerable improvement.

Transfer of introduced DNA to other micro-organisms without making use of its expression for selection will only be easily detectable in a few cases. (1) When the other micro-organism has been defined and can easily be purified. (2) When the rate of transfer is high and when the micro-organism is so predominant that transferred DNA can be detected in randomly chosen colonies.

In these cases the target micro-organism can first be isolated and subsequently screened, using a hybridization assay, for the presence of a specific nucleotide sequence.

When the target micro-organism has not been defined and when it is not a dominant micro-organism, detection at the DNA level can only be done by screening of colonies. Imagine the amount of work necessary to detect one positive non-host bacterium in a total of 10^4 or 10^5 bacteria!

If the introduced DNA is expressed such that colonies containing it can be positively selected, e.g. as antibiotic-resistant or lactose-fermenting clones, the chance of detecting such an event is much higher.

Detection of DNA transfer to other micro-organisms

Techniques described previously for the identification of micro-organisms (e.g. specific antisera and/or outer membrane protein and LPS profiles) can be used to check whether introduced DNA is still present in the "released" bacterium or in another bacterium. When the possibility of transfer to a specific other bacterium (e.g. a related pathogen) should be checked, a large number of individual cells of this bacterium should be first collected (e.g. by "fishing" with antibodies bound to a carrier) and subsequently individual colonies can be checked for the presence of the introduced DNA.

Mechanisms of DNA transfer

DNA can be taken up by a bacterial cell by a number of different mechanisms.

1. Conjugation. This process requires cell-to-cell contact between (at least) two bacterial cells. During the replication process one copy of the (plasmid or chromosomal) DNA is partially or completely transferred from the donor to the acceptor cell. Specific tra(nsfer) genes in the donor are required to enable conjugation.
2. Transduction. This is mediated by a transducing bacteriophage. DNA of such a phage is usually present in the bacterial chromosome in a silent "prophage" stage but it can also be multiplied and yield progeny phages. In addition to their "own" DNA these daughter phages may carry part of the host DNA. This DNA is transferred to a new bacterium as part of the process which occurs when these phages subsequently inject their DNA into a new bacterium.

3. Transformation. This process consists of the uptake of free (circular or linear) DNA by an intact bacterium. The source of free DNA often is a lysed bacterium. Considering the fact that Nature, including bacteria, is rich in DNA-degrading enzymes, free DNA will most probably be degraded. Even when DNA is taken up by a bacterium, the probability of inactivation is high because of the action of the acceptor cell's defence mechanisms of restriction and modification.

METHODS TO REDUCE POSSIBLE RISKS

1. Physical containment. This type of risk reduction can be used in the experimental stage to do research and to do small scale risk assessment. Physical containment of large plots is not feasible.

2. Physiological containment. An essential gene of a released bacterium can be brought under the control of the promoter of another operon in such a way that the released bacterium can only grow under certain conditions, e.g. specific plant products activate the *vir* promoters of *A.tumefaciens* (Okker et al 1984, Stachel et al 1985) and the *nod* promoters of *Rhizobium* (Wijffelman et al 1986, Redmond et al 1986, Peters et al 1986, Firmin et al 1986, Zaat et al 1987). For example, it is likely that strains which produce an essential protein e.g. a RNA polymerase subunit under the control of a flavonoid-activated *nod* promoter can only grow at a certain place at or near the root system of leguminous plants (Redmond et al 1986). Similarly, when an essential gene is brought under the control of the lactose promoter, the strain will only grow as long as lactose is added as a nutrient.

3. Genetic containment. Risks of DNA transfer can be considerably reduced by working with derivatives from which the *tra* gene has been inactivated. Furthermore, strains lacking plasmids are safer than those carrying them. Survival chances of

sporeforming micro-organisms can strongly be reduced by using spore-formation-negative mutants. Bacteria can also be made supersensitive to antibiotics. This property could add an extra safety to initial field trials. However, since such mutations will affect the viability of the strain, it is perhaps more likely that these mutations make the application itself unsuccessful.

4. Genetic time bombs. The principle of this method is that the the released bacterium has been manipulated such that it will destroy itself in due time. If bacteria are made dependent on diaminopimalic acid (dap), an amino acid in the essential peptidoglycan component of rod-shaped bacteria, they lyse upon dap-deprivation. As long as they are supplied with dap, they grofine. Another example is the introduction of a mutation which causes that bacteria do not survive cold, e.g. the winter. The presence of such a mutation is attractive both for venting the spread of the micro-organism as well as for the producer who then can sell a new batch every year. It is also conceivable that bacteria are manipulated such that, under certain conditions, they produce an antibiotic, thereby killing themselves. Finally, it is likely that bacteria can be disabled by preventing the synthesis of certain cell surface components, e.g. flagella (de Weger et al 1987b) or the O-antigen part of LPS (de Weger et al 1987c).

A few comments have to be made on the trend to generate safer strains. Firstly, the remedy may be worse than the disease in some cases, e.g. antibiotic production, since a safety measure may be far worse for the environment than the DNA that was introduced for the purpose of application. Secondly, the construction of strains which are sufficiently stable for applications is often very difficult. Strains containing genetically-engineered DNA are often already weakened as a result of the instability of the construction. Further weakening would make successful application even more unlikely. In several laboratories Rhizobium strains have been constructed which carry multiple copies of the genes required for nodulation. Instead of resulting in more nodules, such strains do not cause nodulation at all! Moreover, under laboratory conditions such constructs can only be maintained under antibiotic selection pressure. The presence of multiple copies decreases their growth rate 7-fold! Several of such examples can be given. It is likely that exchange of genetic material in Nature has taken place very frequently. In many cases this will have resulted in unstable constructions. Those events that so far have resulted in stable constructs are the ones which we presently consider as the "natural" situation.

THE FIRST EXAMPLES OF "DELIBERATELY RELEASED" BACTERIA

Release of genetically-engineered micro-organisms in small plots in the environment has recently started in several places in the world. It is

1. Rhizobium carrying Tn5 has been released in 1987 in Great-Britain, West-Germany and in France in small experimental plots in order to check for gene transfer and to estimate the influence of various plants on numbers of Rhizobia under a range of soil conditions. It should be noted that the Tn5 was transferred to Rhizobium using in vivo techniques, by a process that can occur naturally, i.e. without the help of in vitro recombinant DNA techniques. Therefore this approach is not subject to regulations of Recombinant DNA Advisory Committees in most countries although essentially the same constructs can be made by in vitro genetic manipulation.
2. The "ice-minus" mutant of Pseudomonas syringae has been spread on strawberry plants in the Western part of the United States in 1987.
3. Since Tn5 mutations induced without recombinant DNA techniques were not subject to Dutch Recombinant DNA Advisory Committee regulations in 1985, a field-study could be carried out in the Netherlands in that year to study whether, in contrast to the parental strain, Tn5-induced siderophore-negative mutants of a fluorescent, root-colonizing Pseudomonas strain were unable to reduce potato yield decrease

due to increased cropping frequency. This probably represents the first field experiment with Tn5 mutants. Since 1986 the Dutch Recombinant DNA Advisory Committee follows the policy that their recommendations apply to all Tn5 mutations, no matter by which technique they have been produced.

REFERENCES

Achtman M, Mercer A, Kusecek B, Pohl A, Heuzenroeder M, Aaronson W, Sutton A and Silver RP (1983) Six widespread bacterial clones among Escherichia coli K1 isolates. Infect Immun 39: 315-335

Barry GF (1986) Permanent insertion of foreign genes into the chromosomes of soil bacteria. Bio/Technology 4: 446-449

Becker JO, Hedges RW, Messens E (1986) Diverse effects of some bacterial siderophores on the uptake of iron by plants. In: Swinburne TR (ed) Iron, siderophores, and plant diseases. Plenum, New York. NATO ASI Series A Vol 117 pp. 61-70

Campbell R (1985) Plant microbiology. Edward Arnold, Baltimore p. 184

Chatterjee DK, Kilbarne JJ, Chakrabarty AM (1982) Biodegradation of 2,4,5 trichlorophenoxyacetic acid in soil by a pure culture of Pseudomonas cepacia. Appl Environ Microbiol 44: 514-516

Dijkshoorn L, Michel MF, Degener JE (1987) Cell envelope protein profiles of Acinotobacter calcoaceticus strains isolated in hospitals. J Med Microbiol 23: 313-319

Drahos DJ, Hemming BC, McPherson S (1986) Tracking of recombinant organisms in the environment: β-galactosidase as a selectable non-antibiotic marker for fluorescent Pseudonomads. Bio/Technology 4: 439-444

Fiksel JR, Covello VT (1986) The suitability and applicability of risk assessment methods for environmental applications of biotechnology. In: Fiksel J, Covello VT (eds) Biotechnology risk assessment, issues and methods for environmental introductions. Pergamon Press, New York pp. 1-34

Firmin JL, Wilson KE, Rossen L, Johnston AWB (1986) Flavonoid activation of nodulation genes in Rhizobium reversed by other compounds present in plants. Nature 324: 90-92

Geels FP, Schippers B (1983a) Selection of antagonistic fluorescent Pseudomonas spp. and their root colonization and persistence following treatment of seed potatoes. Phytopathol Z 108: 193-206

Geels FP, Schippers B (1983b) Reduction of yield depressions in high frequency potato cropping soil after seed tuber treatments with antagonistic fluorescent Pseudomonas spp. Phytopath Z 108: 207-214

Goldman RC, Leive L (1980) Heterogeneity in antigenic side chain length in lipopolysaccharide from Escherichia coli O111 and Salmonella typhimurium LT2. Eur J Biochem 107: 145-153

Graham JB, Istock CA (1978) Gene exchange in Bacillus subtilis in soil. Mol Gen Genet 166: 287-290

Graham JB, Istock CA (1979) Gene exchange and natural selection cause Bacillus subtilis to evolve in soil culture. Science 204: 637-639

Graham TL, Watrud LS, Perlak FJ, Tran MT, Lavrick PB, Miller-Wideman MA, Marrone PG, Kaufman RJ (1986) A model genetically engineered pesticide: cloning and expression of the Bacillus thuringiensis subsp. kurstaki δ-endotoxin into Pseudomonas fluorescens. In: Lugtenberg B (ed) Recognition in microbe-plant symbiotic and pathogenic interactions. NATO ASI Series H Vol 4 Springer Verlag, Heidelberg, pp. 385-393

Grant WF (1979) The genotoxic effects of 2,4,5-T. Mutat Res 65: 83-119

Ham GE, Lawn RJ, Brun WA (1976) Influence of inoculation, nitrogen fertilizers and photosynthetic source-sink manipulations on field-grown soybeans. In: Nutman PS (ed) Symbiotic nitrogen fixation in plants. Cambridge University Press, Cambridge, p. 239

Hitchcock PJ, Brown TM (1983) Morphological heterogeneity among Salmonella lipopolysaccharide chemotypes in silver-stained polyacrylamide gels. J Bacteriol 154: 269-277

Hobbs SLA, DeLong CMO, Denes S, Iyer VN (1986) Manipulation of nodulation specificity in the pea Rhizobium leguminosarum symbiosis. In: Lugtenberg B (ed) Recognition in microbe-plant symbiotic and pathogenic interactions. NATO ASI Series H Vol 4 Springer Verlag, Heidelberg pp. 69-78

Hubbard JP, Harman GE, Hadar Y (1983) Effect of soilborne Pseudomonas spp. on the biological control agent, Trichoderma hamatum, on pea seeds. Phytopathology 73: 655-659

Jones RT, Curtiss R (1969) Genetic exchange between Escherichia coli K12 strains in the intestinal tract of mice. Bacteriol Proc 66-67

Kerr A (1980) Biological control of crown gall through production of Agrocin 84. Plant Disease 64: 27-30

Kerr A, Panagopoulos CG (1977) Biotypes of Agrobacterium radiobacter var. tumefaciens and their biological control. Phytopathol Z 90: 172-179

Kloepper JW, Leong J, Teintze M, Schroth MN (1980) Enhanced plant growth by siderophores produced by plant growth-promoting rhizobacteria. Nature 286: 885-886

Kloepper JW, Schroth MN (1981) Relationship of in vitro antibiosis of plant growth-promoting rhizobacteria to plant growth and the displacement of root microflora. Phytopathology 71: 1020-1024

Lambers H (1980) The physiological significance of cyanide resistant respiration in higher plants. Plant cell environ 3: 293-302

Ley P van der, Kuipers O, Tommassen J, Lugtenberg B (1986) O-antigenic chains of lipopolysaccharide prevent binding of antibody molecules to an outer membrane pore protein in Enterobacteriaceae. Microbiol Pathogenesis 1: 43-49

Lewis DM, Bromfield ESP, Barran LR (1987) Effect of rifampin resistance on nodulating competitiveness of Rhizobium meliloti. Can J Microbiol 33: 343-345

Lin W, Okon Y, Hardy RWF (1983) Enhanced mineral uptake by Zea mays and Sorghum bicolor roots inoculated with Azospirillum brasilense. Appl Environ Microbiol 45: 1775-1779

Long SR, Peters NK, Mulligan JT, Dudley ME and Fisher RF (1986) Genetic analysis of Rhizobium-plant interaction. In: Lugtenberg B (ed) Recognition in microbe-plant symbiotic and pathogenic interactions. NATO ASI Series H Vol 4 Springer Verlag, Heidelberg pp. 1-15

Lugtenberg B, Peters R, Bernheimer H, Berendsen W (1976) Influence of cultural conditions and mutations on the composition of the outer membrane proteins of Escherichia coli. Mol Gen Genet 147: 251-262

Lugtenberg B, van Alphen L (1983) Molecular architecture and functioning of the outer membrane of Escherichia coli and other Gram-negative bacteria. Biochim Biophys Act 737: 51-115

Lugtenberg B, van Boxtel R, de Jong M (1984) Atrophic rhinitis in swine: correlation of Pasteurella multocida pathogenicity with membrane protein and lipopolysaccharide patterns. Infect Immun 46: 48-54

Lugtenberg B (1985) Structure and function of outer membrane proteins In: Korhonen TK, Dawes EA, Mäkelä PH (eds) Enterobacteriae surface antigens: methods for molecular characterization. Elsevier Science Publishers, Amsterdam pp. 3-16

Lugtenberg B, van Boxtel R, Evenberg D, de Jong M, Storm P and Frik J (1986) Biochemical and immunological characterization of cell surface proteins of Pasteurella multocida strains causing atrophic rhinitis in swine. Infect Immun 52: 175-182

Marugg JD, van Spanje M, Hoekstra WPM, Schippers B, Weisbeek PJ (1985) Isolation and analysis of genes involved in siderophore-biosynthesis in the plant growth-stimulating Pseudomonas putida strain WCS358. J Bacteriol 164: 563-570

Meyer JM, Abdallah MA (1987) The fluorescent pigment of Pseudomonas fluorescens: biosynthesis, purification and physiochemical properties. J Gen Microbiol 107: 319-328

Miller RV, Saye DJ, Ogunseitan D, Sayler GS (1986) Gene transfer in fresh-water environments. In: Book of abstracts EMBO workshop 'Genetic manipulation of Pseudomonads - Applications in biotechnology and medicine'

Moore LW, Warren G, Strobel GA (1979) Involvement of a plasmid in the hairy root disease of plants caused by Agrobacterium rhizogenes. Plasmid 2: 617-626

Neilands JB (1982) Microbial envelope proteins related to iron. Annu Rev Microbiol 36: 285-309

Neyra CA, Dobereiner J (1977) Denitrification by N2-fixing Spirillum lipoferum. Can J Microbiol 23: 300-305

Nicholas RB (1987) Biotechnology in hazardous-waste disposal: an unfulfilled promise. ASM News 53: 138-142

Noel KD, Brill WJ (1980) Diversity and dynamics of indigenous Rhizobium japonicum populations. Appl Environ Microbiol 40: 931-938

OECD (1986) Recombinant DNA safety considerations

O'Gara F, Treacy P, O'Sullivan M, Higgins P (1986) Biological control of phytopathogens by Pseudomonas spp.: genetic aspects of siderophore production and root colonization. In: Swinburne TR (ed) Iron, siderophores, and plant diseases. NATO ASI Series A Vol 117 Plenum, New York, pp. 331-339

Okker RJH, Spaink H, Hille J, van Brussel AAN, Lugtenberg B and Schilperoort RA (1984) Plant-inducible virulence promoter of the Agrobacterium tumefaciens Ti plasmid. Nature 312: 564-566

Overbeeke N, Lugtenberg B (1980) Major outer membrane proteins of Escherichia coli strains of human origin. J Gen Microbiol 121: 373-380

Palva ET, Mäkelä (1980) Lipopolysaccharide heterogeneity in Salmonella typhimurium analysed by sodium dodecyl sulphate gel electrophoresis. Eur J Biochem 107: 137-143

Panagopoulos CG, Psallidas PG, Alvizatos AS (1979) Evidence for a breakdown in the effectiveness of biological control of crown gall. In: Schippers B, Gams W (eds) Soil borne pathogens. Academic Press, London pp. 569-578

Peters NK, Frost JW, Long SR (1986) A plant flavone, luteolin, induces expression of Rhizobium meliloti nodulation genes. Science 233: 977-980

Pines O, Gutnick DL (1981) Relationship between phage resistance and emulsan production, interaction of phages with the cell surface of Acinetobacter calcoaceticus RAG - 1. Arch Microbiol 130: 129-133

Redmond JW, Batley M, Djordjevic MA, Innes RW, Kuempel PL, Rolfe BG (1986) Flavones induce expression of nodulation genes in Rhizobium. Nature 323: 632-635

Sayler G, Stacey G (1986) Methods for evaluation of microorganism properties. In: Fiksel J, Covello VT (eds) Biotechnology risk assessment. Pergamon Press pp. 35-55

Scher FM (1986) Biological control of Fusarium wilts by Pseudomonas putida and its enhancement by EDDHA. In: Swinburne TR (ed) Iron, siderophores, and plant diseases. NATO ASI Series A, Vol 117 Plenum Press, New York pp. 109-117

Schippers B, Bakker PAHM, Bakker AW, van der Hofstad GAJM, Marugg JD, de Weger LA, Lamers JG, Hoekstra WPM, Lugtenberg BJ and Weisbeek PJ (1986) Molecular aspects of plant growth affecting Pseudomonas species. In: Lugtenberg B (ed) Recognition in microbe-plant symbiotic and pathogenic interactions. NATO Sci Series H, Vol 4, Springer-Verlag, Heidelberg, pp. 395-404

Schnaitman CA (1974) Outer membrane proteins of Escherichia coli IV Differences in outer membrane proteins due to strain and cultural differences. J Bacteriol 118: 454-464

Schneider H, Formal SB, Baron LS (1961) Experimental genetic recombination in vivo between Escherichia coli and Salmonella typhimurium. J Exp Med 114: 141-148

Schroth MN, Hancock J (1983) Disease-suppressive soil and root colonizing bacteria. Science 216: 1376-1382

Stacey G, Brill WJ (1982) Nitrogen fixation by root-inhibiting or infecting bacteria. In: Mount MS, Lacy GH (eds) Phytopathogenic Prokaryotes, Academic Press, New York pp. 225-247

Stachel SE, Van Montagu M, Zambryski P (1985) Identification of the signal molecules produced by wounded plant cells that activate T-DNA transfer in Agrobacterium tumefaciens. Nature 318: 624-629

Strobel GA, Nachmias A (1985) Agrobacterium rhizogenes promotes the initial growth of bare root stock almond. J Gen Microbiol 131: 1245-1249

Swinburne TR (1986) Iron, siderophores, and plant diseases. NATO

ASI Series A, Vol 117 Plenum Press, New York

Trevors JT, Oddie KM (1986) R-plasmid transfer in soil and water. Can J Microbiol 32: 610-613

Weller D, Cook JR (1986) Suppression of root diseases by fluorescent Pseudomonads and mechanism of action. In: Swinburne TR (ed) Iron, siderophores, and plant diseases, NATO ASI Series A, Vol 117 Plenum Press, New York pp. 99-107

Weger LA de, van Boxtel R, van der Burg B, Gruters RA, Geels FP, Schippers B, Lugtenberg B (1986) Siderophores and outer membrane proteins of antagonistic, plant growth-stimulating, root-colonizing Pseudomonas spp. J Bacteriol 165: 585-594

Weger LA de, Jann B, Jann K, Lugtenberg B (1987a) Lipopolysaccharides of Pseudomonas spp. that stimulate plant growth: composition and use for strain identification. J Bacteriol 169: 1441-1446

Weger LA de, van der Vlugt CIM, Bakker PAHM, Schippers B, Lugtenberg B (1987b) Flagella of a plant-growth-stimulating Pseudomonas fluorescens strain are required for colonization of potato roots. J Bacteriol 169: 2769-2773

Weger LA de, Schippers B, Lugtenberg B (1987c) Plant growth stimulation by biological interference in iron metabolism in the rhizosphere. In: Van der Helm D, Neilands J and Winkelman G (eds) Iron transport in microbes and plants and animals, VCH Verlagsgesellschaft pp. 387-400

Weinberg SR, Stotzky G (1972) Conjugation and genetic recombination of Escherichia coli in soil. Soil Biol Biochem 4: 171-189

Wolber PK, Deininger CA, Southworth MW, Vandekerckhove J, van Montagu M, Warren GJ (1986) Identification and purification of ice-nucleation protein. Proc Natl Acad Sci USA 83: 7256-7260

Wijffelman C, Zaat B, Spaink H, Mulders I, van Brussel T, Okker R, Pees E, de Maagd R, Lugtenberg B (1986) Induction of Rhizobium nod genes by flavonoids: differential adaptation of promoter, nodD gene and inducers for various cross-inoculation groups. In: Lugtenberg B (ed) Recognition in microbe-plant symbiotic and pathogenic interactions. NATO ASI Series H, Vol 4 Springer-Verlag, Heidelberg, pp. 123-135

Zaat SAJ, Wijffelman CA, Spaink HP, van Brussel AAN, Okker RJH, Lugtenberg B (1987) Induction of the nodA promoter of Rhizobium leguminosarum Sym plasmid pRL1JI by plant flavanones and flavones. J Bacteriol 169: 198-204

ECOLOGY AND BIOTECHNOLOGY: EXPECTATIONS AND OUTLIERS

Robert K. Colwell
Department of Zoology
University of California
Berkeley, CA 94720

In the absence of retrospective data on risks presented by genetically engineered organisms in the environment, we find ourselves awash in a sea of conjecture. Among biological scientists, there is clearly a broad range of opinions about the likelihood that some organisms produced by molecular and cellular biotechnology may pose unacceptable risks. The distribution of opinion, if it could be quantified, might even prove bimodal.

In this paper, my intention is to explore the reasons for this breadth of expert opinion, in the hope of providing some useful insights. However worthwhile the effort to see the problem through other eyes, though, none of us can fully escape the point of view imposed by his or her own training, experience, and values. I do not pretend be exceptional in this regard. Without doubt, my own background in theoretical and experimental research in evolutionary ecology, and twenty years of work in tropical biology makes it easier for me to understand some points of view than to understand others.

At one extreme of the spectrum of opinion, the most ardent critics of genetic engineering (e.g. Rifkin 1984) see high risk as a certainty and claims of safety as conjectural at best. For those at the other end of the spectrum (e.g. Brill 1985a, 1985b; Davis 1984, 1987), safety appears virtually certain, risks merely conjectural. In fact, whatever the rationale for prediction, the estimation of the level of risk or safety of any hypothetical experiment is necessarily conjectural.

We can distinguish two common approaches to the qualitative assessment of risk in hypothetical experiments or in the implementation of new technologies: theoretical arguments and historical arguments (Fischhoff et al. 1981).

Theoretical Arguments

Theoretical approaches rely on *a priori* estimation of risk on the basis of some conceptual or mathematical model of nature. Theories and models, by their nature, abstract generalities from the mass of particulars that we call reality. As a consequence, the realism of a theory varies inversely with the domain of its applicability (Levins 1966, Colwell 1980)--especially in biology. The more general the theory, the less realistic and precise its predictions for any particular case.

NATO ASI Series, Vol. G18
Safety Assurance for Environmental Introductions
of Genetically-Engineered Organisms
Edited by J. Fiksel and V.T. Covello

Three further problems add to the lack of realism and precision of scientific theories. First, as historians of science point out, prevailing theoretical/conceptual paradigms are often in discord with the preponderance of empirical findings, as scientists cling to "accepted" ideas in the face of mounting contradictions. Second, when scientists in one subfield "borrow" a theory or a guiding concept from a sister field, they often end up with outdated merchandise--not usually completely useless, but at about the level of obsolescence of the average college textbook in general biology. Finally, some soundly discredited ideas seem to have a life of their own, independent of reality, probably because they hold some intrinsic appeal in terms of simplicity, world view, or even political ideals.

In the context of engineered organisms in the environment, theoretical arguments have been widely used to predict average risk, and, consequently, to set the tone of regulatory policy and even to erect *a priori* categories for differing levels of regulatory scrutiny.

The Balance of Nature Paradigm

One key argument rests on the theory of homeostatic balance in natural communities (interacting species) and ecosystems--ecological "buffering." The notion of the "balance of nature" has its roots deep in Western history (MacIntosh 1980, Simberloff 1980). The first evolutionary ecologist, Darwin, clearly saw the balance of nature not as a robust edifice but as a collection of delicately counterpoised elements in continual conflict (Colwell 1985).

In contrast, many ecologists of the early part of this century treated communities as "superorganisms," self-regulated and self-perpetuating, with each species playing a specific role (like the liver or the heart of an animal) for the good of the whole (e.g. Clements 1905). This view of ecosystems produced its most sophisticated descendants in the ecological theories of the 1960's and 70's (e.g. MacArthur 1972), which assume that population interactions (competition, predator/prey, pathogen/host) are normally stable, and that populations are usually at or near their theoretical equilibrium.

Consciously or unconsciously, the image of resilient, self-regulating biological communities and ecosystems may have special appeal to scientists who work daily with cellular or organismal homeostatic mechanisms--the narrow bounds of calcium balance within cells, or the pH of blood.

In arguments about risk from genetically organisms, the balance-of-nature paradigm has frequently been called upon to support the argument that novel genotypes are not likely to prosper in natural (or agricultural) environments, and thus present little or no risk, because "nature is resilient," "buffered," its balance not easily upset (e.g. Bentley 1984; Brill 1985a, 1985b; Regal 1986 reviews this issue). Certainly there are homeostatic forces in communities and ecosystems, but we now know that the limits of stability are often narrow, and that once those limits are passed, positive feedback forces frequently take over, exacerbating problems. The eutrophication of lakes, epidemics of plant pathogens in crop monocultures (Doyle 1985), or the spread of introduced species (Mack 1981) represent three fairly typical, well-studied examples.

Ecologists have increasingly recognized that many, perhaps all, communities are much more often out of ecological equilibrium than near it (Sousa 1984, Colwell 1985, Pickett and White 1985). This discovery implies that communities are far more invasible than previously believed (Simberloff and Colwell 1984)--a theme I will return to later.

The Niche Paradigm

Closely related to the balance-of-nature paradigm, both historically and logically, is the notion of "pre-existing niches." The niche concept was introduced by vertebrate biologists in the early part of this century to describe the differences in ecological role of animals that live together in the same habitat. By the late 1950's, the term had become clearly defined in theoretical ecology as a purely functional concept, entirely separate from ideas of physical space (habitat or microhabitat) (Hutchinson 1957, Colwell and Fuentes 1975). The business world has borrowed the term from ecology for an analogous, functional use--a specialized market for goods or services. Microbial ecologists use the term not only in the functional sense, but continue to include a vague spatial sense as well.

The idea that *functional* niches exist, in some sense, independent of the species that "occupy" them, presents serious philosophical problems for a positivist science, and serious practical ones for the hypothetico-deductive method. Nowadays, most theoretical ecologists are careful to define "niche" as a *property of* a particular population or species--essentially an extended phenotype (Colwell and Fuentes 1975).

To say that the niche of a hummingbird species in Mexico and the niche of a sunbird species in southern Africa (unrelated species) are remarkably similar, in comparison with the similarity between the niches of hummingbirds and warblers, is a meaningful and testable scientific statement. (Hummingbirds are found only in the New World, sunbirds only in the Old World.) To say that the hummingbird and the sunbird "occupy the same niche on different continents," on the other hand, is not only untestable, but grossly misleading. If the hummingbird became extinct, the resources thus freed up might just as well be partitioned among bees (which take nectar) and insectivorous birds (some of which, like hummingbirds, feed on aerial insects) as taken over by another hummingbird-analogue. To make the point another way: Is the "hummingbird niche" in Antarctica "empty," or non-existent?

In nature, resources (habitats, substrates, nutrients, food types) not currently fully utilized (like unsaturated markets in the business world) frequently exist, but the means by which they are eventually exploited depends on history, context, and a large measure of happenstance. Furthermore, resources currently utilized inefficiently, or defended ineffectively by one species are just as likely to be partitioned among two or more species with somewhat different niches as to be expropriated wholesale by a single stronger competitor with very similar niche characteristics. One-for-one substitutability of one species by an analogous one from another community remains--for lack of knowledge--an open question for microbes, but it is fully discredited for higher organisms.

Once we abandon the scientifically unsupportable notion of specific, pre-existing niches, implications for the fate of engineered organisms in the environment become clear. Consider, for a start, the argument that naturally occurring species (or strains) are "optimally adapted to their (pre-existing) niche," and therefore are likely to outcompete any genetically modified version of the same species (e.g. Davis 1984, 1987; and Brock 1986; see also Colwell 1986, Regal 1986, and Sharples 1987). It may be true that some species (especially microbes in fairly stable microhabitats) are indeed optimally or near-optimally adapted to the current conditions of their environment, *given* the current genetic potential of the species. Except when genetic changes produced in the laboratory duplicate or very closely mimic naturally-occurring genotypes, however, the fate of the introduction of the novel genotype cannot be predicted with any confidence on the basis of this optimality argument. There is no pre-existing niche in which it must either "fit" or perish--only pre-existing resources which may (or may not) be exploited in some novel way or in a novel combination.

Moreover, once we abandon the Platonic world of pre-existing niches, each optimally filled with an occupying species, the concept of a fully-saturated community that is resistant to invasion by novel genotypes seems far less plausible. We will take a closer look at this issue in a later section.

The Proportionality Paradigm

The argument--or more often, the unstated assumption--that small genetic changes necessarily produce only small ecological changes lies at the heart of many *a priori* arguments for the presumption of safety, and provides the basis for current proposals for regulatory guidelines in the U. S. A. (Office of Science and Technology Policy 1986). The rationale for this paradigm of "proportionality" between degree of genetic change and degree of ecological change comes in part from the notion of pre-existing niches: unless some major change is introduced into the genotype, the modified organism will still "occupy the same niche" in the environment that parent organism occupied.

This argument often makes sense, when restated in testable terms. Indeed, in many cases--perhaps most, the addition or deletion of a single gene may be insufficient to release the novel organism from the potential for direct competition for resources with its parent genotype. It makes sense, however *only so long as the new genotype remains in the microhabitat of its parent genotype* (we will return to this point presently). Even so, there is no theoretical guarantee that the new genotype will be less fit than the parent (and therefore safe), although even conscious design for survival by human experimenters will probably fail more often than it succeeds in competition with the products of 3 billion years of evolution.

To complicate matters, we now know for higher organisms that continuous, intense competition between similar species in nature is far from universal (e.g. Connell 1983, Schoener 1983, Strong et al. 1984). This finding is directly related to the discovery that natural communities are often not in ecological equilibrium, due to disturbance, heterogeneities in space and time, or to the intermittent action of natural enemies.

It remains to be seen to what degree these findings apply to microbial communities. Competition is perhaps more common among microbes than among higher organisms, although the appropriate experiments with microbes outside the laboratory have been relatively few. Even with microbial species, however, the colonization of newly available patches of unoccupied substrate (newly opened leaves or flowers, newborn animals, lesions, newly exposed organic matter) often shows a strong "priority effect" or "pre-emptive competition." Among two strains that are approximately equal in competitive potential, the first to arrive and reach population equilibrium may exclude the other, in spite of genetic differences (Atlas and Bartha 1987). This pattern was demonstrated, for example, for ice-minus deletion mutants (naturally-occurring or engineered) of *Pseudomonas syringae* in competition with the corresponding parent strains (S. Lindow, personal communication).

The qualification that competition between a genetically engineered organism and its parent genotype is more likely when the new genotype *remains in the microhabitat of its parent genotype* holds the key to one of the principal conjectural risks from genetically engineered organisms. All organisms are geographically limited by climatic factors, and within their potential geographic range, they are further limited in distribution to particular habitats or microhabitats by biotic or abiotic factors. Microbes are no exception (e.g. Atlas and Bartha 1987, Sayler and Stacy 1986, Levy 1986b).

The concern lies in the fact that many of the small genetic changes proposed for engineered organisms involve an intentional expansion of ecological range. Crop plants, for example, are being engineered to grow in colder climates; in dryer, saltier, or nitrogen-poor soils; and to resist attack by pathogens and arthropod pests. All of these factors now limit geographical ranges (Elkington 1986). Microbes are being developed to grow at higher temperatures (Sayler and Stacy 1986), and to colonize novel hosts (Betz et al. 1983), for purposes of broader and more effective biological control of pests (e.g. by expanded host range and virulence in *Bacillus thuringiensis*). Host range shifts can often be produced by the simplest change of all, the deletion of an appropriate gene (e.g. Nester et al. 1986).

These kinds of ecological shifts, if successful, may have at least three important effects. First they release the engineered genotype from the potential for head-to-head competition with the parent genotype. Second, they put the new genotype into contact with an expanded or altered range of other species, some of which may never have encountered the parent genotype. Third, such shifts may have direct or indirect effects on the habitat itself. All three phenomena are cause for close scrutiny of unanticipated risks (Colwell et al. 1985, 1987; Sharples 1987).

Examples of the first effect are not hard to find among fungi, bacteria, and viruses. Whenever mutations arise that permit reproduction on previously resistant hosts, the mutant strain has escaped from competition with the non-mutant (parent) strain. Familiar examples include the evolution of antibiotic resistance (Levy 1986a, 1986b) and the evolutionary races between pathogens and plant breeders (Doyle 1985) and between viral strains and immunologists (Palese 1986)--frequently based on very small and very simple genetic changes.

The history of "cheatgrass" (*Bromus tectorum*, official common name "downy brome") in the western United States may be an example of a natural genetic change and consequent ecological shift that demonstrates the second and third effects--contact with an expanded number of other species and effects on habitat. Cheatgrass arrived from Europe in the last century and spread along trails and roads into overgrazed areas in the northern Great Basin, where it eventually replaced more valuable native grasses over immense areas of land, finally reaching a stable range about 1930 (Mack 1981). Recently, however, the grass has begun a rapid secondary spread into drier rangeland than it previously occupied, where it becomes tinder for wildfires that destroy surrounding native vegetation and valuable forage. Although changes in grazing regimen are probably also involved, Young (1987) suggests that this secondary spread may have been facilitated by natural selection for drought tolerance.

A special form of the "proportionality paradigm" is the concept that the placement of genetic regulatory sequences ("noncoding regulatory regions") in novel genetic contexts will prove innocuous, because--in themselves--regulatory sequences produce no gene products, no "new traits" (Office of Science and Technology Policy 1986). In the laboratory, this argument may appear perfectly reasonable. Focusing on some particular gene product of interest, the molecular biologist inserts a small piece of DNA that causes the organism to produce much more of that desired product--a quantitative, not a qualitative change in the phenotype. Changes in gene regulation, however, do not always produce simple changes, but sometimes have pleiotropic (multiple) effects on different aspects of phenotype (Paigen 1986), some of which might be not be expressed under laboratory growth conditions, or might not be noticed unless specifically sought.

More dubious still is the next step in the logic--that a quantitative phenotypic change will necessarily yield a simple, proportional change in ecological relations when the organism is introduced into the environment in large numbers (Colwell et al. 1987). Moderate changes in growth rate (e.g Andrews and Hegeman 1976) or in virulence (Anderson and May 1979, May and Anderson 1979) may profoundly alter not only quantitative but also the qualitative relations between species. Without exception, mathematical models of species interactions predict strongly nonlinear behavior--thresholds, accelerations, limit cycles, and "neighborhood" stability--and evidence from field studies confirms many of these predictions (Anderson and May 1978, May and Anderson 1978, May 1981).

In summary, the idea that small genetic changes *necessarily* yield small ecological changes is not supportable, even though likely to be true in many cases.

Arguments from Historical Precedents

Apparently in agreement with the historian Trevelyan that "several imperfect readings of history are better than none at all," molecular biologists, geneticists, plant and animal breeders, plant pathologists, and ecologists have all brought forward case histories, anecdotes, and statistics intended to bear on the question of risk assessment for genetically engineered organisms in the environment.

Unfortunately, instead of using these diverse versions of history to triangulate on the future, we have sometimes spent our energies in either denying the relevance of certain antecedents, or even denying their veracity. It is trivially true that no historical precedent precisely fits a future prospect--which makes all precedents easy targets for unsympathetic skeptics.

The Regulatory History of the RAC

The very successful history of "self-regulation" of recombinant DNA research in the laboratory, under the aegis of the Recombinant DNA Advisory Committee in the United States, provides an excellent model for flexible regulation based on increasing experience with a new technology. Some influential molecular and microbial biologists have argued that the reassuring record of safety with recombinant organisms in laboratory research provided evidence of safety for organisms designed for environmental use (e.g. Davis 1987).

By regulation, however, most laboratory research before 1980 was conducted with strains *designed to perish* outside special conditions provided in the laboratory; moreover, risk assessment was clearly focused on human hazards (Zimmerman 1984, Levy 1986b). The fact that some of these organisms undoubtedly did leave the laboratory, but caused no harm, is certainly reassuring. Unfortunately, however, this experience has limited bearing on ecological risks presented by engineered organisms *designed to survive and reproduce* outside the laboratory. Moreover, the RAC itself, at the time of its first approvals of genetically engineered organisms for environmental testing, in 1982-3, was ill-equipped to assess potential ecological effects of these tests; not a single professional ecologist or evolutionary biologist sat on the RAC at the time (Colwell 1987).

The history of the RAC, then, while providing an excellent historical model for flexibility in regulation, cannot in itself answer questions of risk for genetically-engineered organisms in the environment.

Traditional (Organismal) Biotechnology

As we all hasten to point out to the surprising number of people who seem to take ethical offense at the notion of "altering genes" in microbes and, especially, in higher organisms, human beings have been carrying out biotechnology at the organismal level since prehistory. In recent decades, classical mutation/selection and hybridization techniques have even succeeded in moving desirable single genes or polygenic traits from one strain or species into another (e.g. Goodman et al. 1987).

Many have argued that molecular and cellular biotechnology (to use Hardy's [1985] terminology) is simply a better way to do the same things we have done for years with organismal biotechnology, and therefore that the products of these new techniques require no new efforts at risk assessment (e.g. Brill 1985a, 1985b). To be sure, there is relevance and reassurance in the history of environmental safety in the development and use of microbial pesticides, commercial *Rhizobium* strains, most biological control agents, most live vaccines, nearly all cultivated plants, and the majority of domesticated animals. In each of these categories, the detailed experience of experts is invaluable, not only in developing safe organisms, but in designing effective ones.

On the other hand, far from all proposed genetic amendments to existing organisms could even in theory have been carried out with classical techniques. For example, the first "intergeneric" microbial pesticide to be considered by the U.S. Environmental Protection Agency was produced (by Monsanto) by inserting the toxin-producing gene from the leaf-dwelling bacterium *Bacillus thuringiensis* (BT) into the cosmopolitan soil bacterium *Pseudomonas fluorescens*. The Monsanto organism is clearly something new under the sun--there is little or no evidence for gene exchange between gram-positive and gram-negative bacteria. The difference in microhabitat decreases further the probability that Nature has already tried this combination and found it wanting, although decaying leaves certainly bring leaf bacteria in contact with the soil.

Although *Bacillus thuringiensis* itself is a registered pesticide, the combination poses many new questions. The fact that an organism could not have arisen by natural processes certainly does not guarantee that it will cause ecological problems. Nonetheless, the novel aspects of the organism are obviously a reasonable focus for risk assessment. In the Monsanto case, the primary concern might be called a question of packaging and delivery. BT lives on leaf surfaces, where caterpillars eat it with leaf tissue, get sick, and die. The Monsanto organism lives in the soil, in a very different ecological community.

The company's greenhouse tests show that their organism effectively kills the target pest--root-feeding moth larvae--"cutworms," a serious pest of corn that currently must be controlled with chemical pesticides. Because the BT toxin is not usually "delivered" in the soil, however, the complex community of soil organisms that maintain soil texture and fertility must be considered for adverse effects. Moreover, runoff into streams may be expected, so the stream community is potentially at risk as well. And of course, the potential for harm to beneficial relatives of the target pest must also be examined--butterflies, in addition to being objects of our admiration, are important pollinators of many native plants. Monsanto has undertaken testing for all these issues.

Long experience with the risks and benefits of organismal technology provides an appropriate and sensible baseline for evaluating the products of molecular and cellular biotechnology, but additional ecological questions may arise with genetic constructions not previously possible.

Introduced Non-native Organisms

In discussions of ecological risk assessment for engineered organisms, ecologists persist in raising the issue of historical introductions of non-native organisms, pointing out the ecological dislocations--and sometimes genuine disasters--that they have caused (Sharples 1983, 1986, 1987; Brown et al. 1984; Levin and Harwell 1984; Simberloff and Colwell 1984; Colwell et al. 1985; Regal 1986). Whether or not these anecdotes and statistics are relevant to risk assessment of genetically engineered organisms has been hotly debated (e.g. Davis 1985, 1987; Simberloff and Colwell 1984; Regal 1986). As with the other historical precedents that I have discussed, the answer appears to be "Yes, and No."

Objections to using introduced non-natives as a model for the introduction of genetically engineered organisms center on two points. Most ecologists would fully accept the first, if it is properly qualified: that minor genetic modifications of indigenous species are a poor fit to the introduced non-natives model. The qualification required is that such a species *must remain in the habitat (or microhabitat) of the parent genotype.* On the other hand, genetic changes that broaden or shift the geographic range of a species, as discussed above, are a different matter, and may indeed be a legitimate parallel to introduced non-native species.

The second objection is also partly correct--that the spread of non-native pests occurs because their natural enemies and competitors are absent in their new home (e.g. Davis 1984, 1987). The statement is true for some cases, but not by any means for all. In the case of introduced predators and pathogens, for example, the problems are usually caused by the absence of appropriate, co-evolved defenses in victim species (the problem in the case chestnut blight, or of the Nile perch, discussed below). This phenomenon provides a valid model for risks to non-target species in the application of genetically engineered microbial pesticides and other agents of biological control (e.g. engineered herbivorous insects for the control of weeds, when the technology becomes available).

In addition, the belief that organisms become pests only in the absence of natural controls is clearly contradicted by one large class of evidence. If problems arise only for this reason, then *native* species should rarely become pests, because the normal controls remain in place. In fact, "native pests" are exceedingly common. Pimentel (1986a) has computed that 27% of the important weeds of U.S. crops are native species; the figure is 59% for pasture weeds, 60% for pest arthropods of crops, 73% for pest arthropods of forests, and 29% for pests of livestock; 30% of all major plant pathogens in the U. S. are natives.

Some of these pests may indeed have escaped from natural control through the destruction of native natural enemies that have been inadvertently eliminated by pesticides or other means, but many cases do not fit this pattern. There is reason to suspect that genetic changes--perhaps small ones, as in the case of the apple maggot (Bush 1974)--have often facilitated the initiation into pesthood of these native species. Further study on this question is urgently needed, and could be aided by molecular techniques.

Recent efforts have made available several comprehensive and up-to-date surveys and analytical studies of introduced non-native species (Mooney and Drake 1986, Groves and Burden 1986, Kornberg and Williamson 1987, and Sharples 1987). Some general patterns have begun to emerge from these studies (Case 1987). For example, it appears that, on average, an indigenous natural enemy (herbivores, predators, parasitoids, pathogens) can deal better with a non-native victim (food plants, prey, and hosts, respectively) than indigenous victims can deal with non-native enemies--a formalization of the patterns discussed above.

In addition to providing some guidance on what to look for in assessing risks for engineered organisms, the history of introduction of non-native organisms teaches a painful lesson on the value of risk assessment itself: careful prior study of particular cases could nearly always have predicted the problems that later arose. Here, of course, I speak of intentional introductions--the best parallel to the intentional introduction of a genetically engineered organism designed to reproduce outside the habitat of its parent organism.

Hindsight, of course, is notoriously acute, but the history of intentional introductions of non-native organisms for purposes of biological control, in which careful prior study *is* a standard practice, provides a strong (though not perfect [Howarth 1983]) record of safety (Pimentel 1986b). In contrast, case after case of well-meaning introductions, often carried out under official government sanction, have lead to "unexpected" results that would have been predicted by any appropriately constituted group of specialists in the natural history--and, in some cases, the physiology, biochemistry, and genetics of the parent organism.

Williams (1980), for example, outlines a risk assessment strategy for non-native species under consideration for introduction as forage plants, based on an evaluation of the characteristics of 36 serious weeds (including the notorious kudzu, *Pueraria lobata*) that were introduced to the U. S. intentionally for use as forage crops, ornamentals, and other objectives. In a lesson for the careful planning of mitigation, Williams (1980) relates that a candidate for use as a new forage plant (goatsrue, *Galega officinalis*) was tested at an experimental farm, rejected as unpalatable to livestock, but not fully extirpated from the test plots. The plant escaped and has become a serious, though still local problem, spreading north from the test site.

As a truly disastrous example of needlessly poor risk assessment, consider the case of the Nile perch (*Lates niloticus*), stocked since 1960 in Lake Victoria (where is not native) in East Africa as a food source for the local populace. This large, predatory fish led to the disappearance of literally hundreds of endemic fish species in the lake; the Nile perch now feeds on small shrimp and its own young (Hughes 1986). But the effects of this introduction do not stop at the lakeshore. Fish have always been a staple food of people living near the lake. The Nile perch is edible, but requires a higher cooking temperature than the native fish did, to render its fat. The consequent effects of intensified firewood cutting on local forests may eventually prove a greater economic and ecological disaster than the loss of the native fish fauna.

As in many other cases, none of the facts that would have been required to have predicted this scenario would have been hard to discover--apparently there simply was no effort to do so. Never previously exposed to any predator as large as the Nile perch, the local fish species were defenseless--a fact easily determined by experiments with captive fish. Prior determination of the cooking temperature of Nile perch vs. native fish would have been a simple matter.

In spite of the principles and caveats they reveal, *particular* case histories of problems caused by introduced non-native species, in themselves, are probably of little direct value. Furthermore, like other phenomena that involve a degree of risk, the hazards presented by the introduction of non-native organisms are not all-or-none, but cover a continuum of severity. The continuum starts at zero (for the great but uncounted majority of introductions that never become established, and many more that establish themselves but cause no harm) and ends with the catastrophic outliers, like chestnut blight, kudzu, or the Nile perch. Although "risk outliers" such as these represent a minuscule proportion of the the total number of species introductions, economically and in terms of public perception, they are serious indeed.

Expectations and Outliers

Both theory and history provide some guidance in risk assessment for genetically engineered organisms, yet even the best mixture of insight and experience one can distil from these considerations does not permit the design of generic guidelines at the level of generality accepted for the testing of new drugs or the toxicity of chemicals.

The problem of risk in biotechnology--both old and new, is not a problem of averages, but of question of outliers. It is my conjecture that the *average* risk to the environment from the products of traditional (organismal) biotechnology and from the products of genetic engineering will prove to be about equal, for any given kind of product. On the other hand, it is possible that the *variability* in risk among successive individual cases many well be higher for genetically engineered organisms, producing a broader risk profile (Fiksel and Covello 1986), as suggested in Figure 1.

The rationale for this possibility is twofold. At the low-risk end of the spectrum, as many have pointed out (e.g. Brill 1985a, 1985b; Davis 1987), the exacting tools of molecular biotechnology permit more precise and better-characterized genetic modifications than traditional mutation/selection (organismal) techniques. This greater degree of precision and repeatability may in itself promote safety and predictability, but in addition, it provides the basis for more accurate monitoring of field tests, as well as the potential to incorporate safety measures in new strains of organisms that are not feasible by traditional techniques.

Balancing these lower-risk cases is the potential for higher-risk outliers. If not detected by risk assessment before release, these will be the "ecological surprises," paralleling events like the sudden spread of chestnut blight or cheatgrass, the depredations of the Nile perch, or the shift of a native insect from a wild plant to a cultivated one.

Why should molecular techniques have a greater potential for producing ecological surprises than traditional mutation/selection techniques? Because recombinant techniques have the ability to combine the independent "evolutionary inventions" of biological lineages that do not exchange genetic information in nature, or do so with a negligible frequency (Krieber and Rose 1986). This ability provides a greater potential for the production of ecological novelty than it is usually possible to achieve with classical techniques (Regal 1986, Colwell et al. 1987).

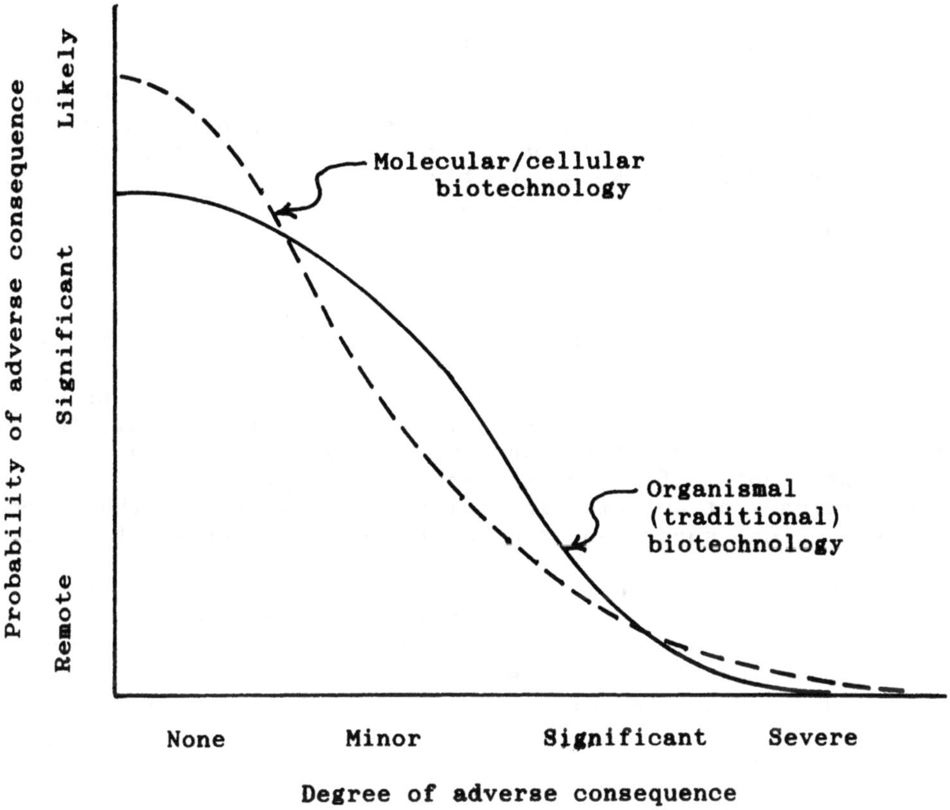

Figure 1. A hypothetical risk profile for the products of organismal biotechnology (traditional mutation/selection techniques) vs. comparable products of modern molecular and cellular biotechnology, intended for environmental use. Units of the ordinate might be scaled as probability per product. Units of the abscissa would necessarily vary with the kind of hazard considered. Both scales should be considered logarithmic.

It must be said that this aspect of biotechnology seems to be consistently downplayed by many of its practitioners (e.g. Davis 1987), in contrast with the concerns of many ecologists and evolutionary biologists (Colwell et al. 1987). In addition to the obvious explanation provided by the difference in professional focus of these two groups of scientists, I believe there may be a simpler explanation. Many of the key tools of modern molecular biology, with which the molecular biologist works daily, are derived from natural means of interspecific genetic transfer--viruses, plasmids, *Agrobacterium*, and so on; perhaps this fact places these mechanisms front and center in the mental theater. Yet in nature, these mechanisms operate under extremely restrictive circumstances (*de novo* intense selection) and among relatively narrow groups of organisms--far narrower than biotechnology hopes they will remain in human hands.

Certainly most engineered organisms will not combine adaptations derived from distantly related evolutionary lineages, and probably most of those that do will cause no harm, if they survive at all. Outliers, by definition, lie beyond the mass of cases. The absolute *number* of outliers in any distribution, however, depends not only on the variance of the distribution but on the sample size. No one doubts that the pace of development of experimental genotypes will accelerate with expanding use of molecular techniques--the "sampling rate" of possibilities is increasing.

One source of disagreement about ecological risk in biotechnology may stem in part from conflicting philosophical views about outliers. Molecular biologists are reassured by the highly deterministic control that is now possible over precise genetic constructions, which in fact may provide a basis for low-risk outliers. Ecologists and evolutionary biologists, in contrast, tend to focus on the inherent complexity and indeterminacy of outcomes in biological communities--the source of "ecological surprises" that characterize adverse outliers.

Whatever the true risk profile may be, the goal of risk assessment must be to detect extremes of both kinds. The production of low-risk organisms should be encouraged in two ways: by expediting their passage through the regulatory apparatus, and by the screening out of high-risk outliers.

Ironically, in parallel with the model of combining genes from lineages long independent in their evolution, this decade has, for the first time, brought scientists together in the public policy arena from opposite ends of the biological hierarchy that leads from molecules to ecosystems--after some 20 years of relatively independent evolution. As a secondary consequence, microbial ecologists and those of us that primarily study the ecology of higher organisms have been forced to attempt convergence, out of our own separate evolutionary trajectories within the field of ecology (Colwell 1986). These crossings have not always proceeded easily. In my view, the best hope for the development of a fair, efficient, and effective system of risk assessment lies in acknowledging--and respecting--our different views of history, the inherently distinct levels of determinacy in the systems we study, and the equivalent value of our realms of study to science and humanity.

References Cited

Anderson, R. M., and R. M. May. 1978. Regulation and stability of host-parasite population interactions: I, Regulatory processes. Journal of Animal Ecology 47:219-247.

Anderson, R. M., and R. M. May. 1979. The population biology of infectious diseases: Part I. Nature 280:361-367.

Andrews, K. J., and G. D. Hegeman. 1976. Selective disadvantage of nonfunctional protein synthesis in *Escherichia coli*. Journal of Molecular Evolution 8:317-28.

Atlas, R. M., and R. Bartha. 1987. Microbial ecology: fundamentals and applications. Benjamin/Cummings, Menlo Park, Calif.

Bentley, O. G. 1984. Statement before U. S. House Committee on Energy and Commerce, Subcommittee on Oversight and Investigations, December 11. U. S. Department of Agriculture, Washington, D. C. (Cited in Fiksel and Covello 1986).

Betz, F., M. Levin, and M. Rogul. 1986. Safety aspects of genetically-engineered microbial pesticides. Recombinant DNA Technical Bulletin 6:135-141.

Brill, W. J. 1985a. Safety concerns and genetic engineering in agriculture. Science 227:381-384.

Brill, W. J. 1985b. Genetic engineering in agriculture. Science 229:115-118.

Brock, T. D. 1986. Prokaryotic population ecology. Pp. 176-179 *in* Halvorson, H. O., D. Pramer, and M. Rogul (eds.), Engineered organisms in the environment: scientific issues. American Society for Microbiology, Washington, D. C.

Brown, J. H., R. K. Colwell, R. E. Lenski, B. R. Levin, M. Lloyd, P. J. Regal, and D. Simberloff. 1984. Report on workshop on possible ecological and evolutionary impacts of bioengineered organisms released into the environment. Bulletin of the Ecological Society of America 65: 436-438.

Bush, G. L. 1974. The mechanism of sympatric host race formation in the true fruit flies (Tephritidae). Pp. 3-23 *in* White, M. J. D. (ed.), Genetic analysis of speciation mechanisms. Australia and New Zealand Book Co., Sydney.

Case, T. J. 1987. Travelers and their fate. Science 236:1000-1002.

Colwell, R. K. 1984. What's new? Community ecology discovers biology. Pp. 387-397 *in* Price, P. W., C. N. Slobodchikoff, and W. S. Gaud (eds.), A new ecology: novel approaches to interactive systems. Wiley, N. Y.

Colwell, R. K. 1985. The evolution of ecology. American Zoologist 25: 771-777.

Colwell, R. K. 1986. Biological responses to perturbation: Genome to ecosystem. Pp. 230-232 *in* Halvorson, H. O., D. Pramer, and M. Rogul (eds.), Engineered organisms in the environment: scientific issues. American Society for Microbiology, Washington, D. C.

Colwell, R. K. 1987. The ice-minus case and a scientifically informed judiciary. Science 237:10.

Colwell, R. K., and E. R. Fuentes. 1975. Experimental studies of the niche. Annual Review of Ecology and Systematics 6: 281-310.

Colwell, R. K., E. A. Norse, D. Pimentel, F. E. Sharples, and D. Simberloff. 1985. Genetic engineering in agriculture. Science 229: 111-112.

Colwell, R. K., L. W. Barnthouse, A. Dobson, F. Taub, and R. Wetzler. 1987. Response to the Office of Science and Technology policy notice "Coordinated framework for regulation of biotechnology." Bulletin of the Ecological Society of America 68:16-23.

Connell, J. H. 1983. On the prevalence and relative importance of interspecific competition. American Naturalist 122:661-696.

Clements, F. E. 1905. Research methods in ecology. University Publishing Co., Lincoln, Nebr.
Davis, B. D. 1984. Science, fanaticism, and the law. Genetic Engineering News 4(5):4.
Davis, B. D. 1987. Bacterial domestication: underlying assumptions. Science 235:1329-1335.
Doyle, Jack. 1985. Altered harvest. Viking, N.Y.
Elkington, J. 1986. Double dividends? U. S. Biotechnology and Third World development. World Resources Institute, Washington, D. C.
Fiksel, J., and V. T. Covello. 1986. The suitability and applicability of risk assessment methods for environmental applications of biotechnology. Pp. 1-34 *in* Fiksel, J., and V. T. Covello (eds.), Biotechnology risk assessment: issues and methods for environmental introductions. Pergamon, N.Y.
Fischhoff, B., S. Lichtenstein, P. Slovic, S. L. Derby, and R. L. Keeney. 1981. Acceptable risk. Cambridge University Press, Cambridge.
Goodman, R. M., H. Hauptli, A. Crossway, and V. C. Knauf. 1987. Gene transfer in crop improvement. Science 236:48-54.
Groves, R. H., and J. J. Burden (eds.). 1986. Ecology of biological invasions. Cambridge Univ. Press, N.Y.
Hardy, R. W. F. 1985. Biotechnology in agriculture: status, potential, concerns. Pp. 99-113 *in* Teich, A. H., M. A. Levin, and J. H. Pace (eds.), Biotechnology and the environment: risk and regulation. American Association for the Advancement of Science, Washington, D. C.
Howarth, F. G. 1983. Classical biocontrol: panacea or Pandora's Box. Proceedings of the Hawaiian Entomological Society 24:239-244.
Hughes, N. F. 1986. Changes in the feeding biology of the Nile perch, *Lates niloticus* (L.) (Pisces: Centropomidae), in Lake Victoria, East Africa, since its introduction in 1960, and its impact on the native fish community of the Nyanza Gulf. Journal of Fisheries Biology 29:541-548.
Hutchinson, G. E. 1957. Concluding remarks. Cold Spring Harbor Symposium on Quantitative Biology 22:415-427.
Kornberg, H., and M. H. Williamson (eds.). 1987. Quantitative aspects of the ecology of biological invasions. Royal Society, London.
Krieber, M., and M. R. Rose. 1986. Molecular aspects of the species barrier. Annual Review of Ecology and Systematics 17:465-485.
Levin, S. A., and M. A. Harwell. 1984. Potential ecological consequences of genetically engineered organisms. Pp. 104-136 *in* Gillett, J. W., S. A. Levin, M. A. Harwell, M. Alexander, D. A. Andow, and A. M. Stern (eds.), Potential impacts of environmental release of biotechnology products: assessment, regulation, and research needs. Ecosystems Research Center, Cornell Univ., Ithaca, N.Y.
Levins, R. 1966. The strategy of model building in population biology. American Scientist 54:421-431.
Levy, S. B. 1986a. Ecology of plasmids and unique DNA sequences. Pp. 180-190 *in* Halvorson, H. O., D. Pramer, and M. Rogul (eds.), Engineered organisms in the environment: scientific issues. American Society for Microbiology, Washington, D. C.

Levy, S. B. 1986b. Human exposure and effects analysis for genetically modified bacteria. Pp. 56-74 *in* Fiksel, J., and V. T. Covello (eds.), Biotechnology risk assessment: issues and methods for environmental introductions. Pergamon, N Y.

MacArthur, R. H. 1972. Geographical ecology: patterns in the distribution of species. Harper and Row, N.Y.

MacIntosh, R. P. 1980. The background and some current problems of theoretical ecology. Synthese 43:195-255.

Mack, R. N. 1981. Invasion of *Bromus tectorum* L. into western North America: an ecological chronicle. Agro-Ecosystems 7:145-165.

May, R. M. 1981. Theoretical ecology, 2nd ed. Blackwell Scientific Publications, Oxford.

May, R. M., and R. M. Anderson. 1978. Regulation and stability of host-parasite population interactions: II, Destabilizing processes. Journal of Animal Ecology 47:249-267.

May, R. M., and R. M. Anderson. 1979. The population biology of infectious diseases: Part II. Nature 280:455-461.

Mooney, H. A., and J. A. Drake (eds.). 1986. Ecology of biological invasions of North America and Hawaii. Ecological Studies 58. Springer-Verlag, N Y.

Nester, E. W., M. F. Yanofsky, and M. P. Gordon. 1986. Molecular analysis of the host range of *Agrobacterium tumefaciens*. Pp. 191-196 *in* Halvorson, H. O., D. Pramer, and M. Rogul (eds.), Engineered organisms in the environment: scientific issues. American Society for Microbiology, Washington, D. C.

Office of Science and Technology Policy, U.S. 1986. Coordinated framework for regulation of biotechnology. Federal Register 51:23301-23350.

Paigen, K. 1986. Gene regulation and its role in evolutionary processes. Pp. 3-36 *in* Karlin, S., and E. Nevo, Evolutionary processes and theory. Orlando, Fla.

Palese, P. 1986. Rapid evolution of human influenza viruses. Pp. 53-68 *in* Karlin, S., and E. Nevo, Evolutionary processes and theory. Orlando, Fla.

Pickett, S. T. A., and P. S. White. 1985. The ecology of natural disturbance. Academic Press, Orlando, Fla.

Pimentel, D. 1986a. Biological invasions of plants and animals in agriculture and forestry. Pp. 149-162 *in* Mooney, H. A., and J. A. Drake, Ecology of biological invasions of North America and Hawaii. Ecological Studies 58. Springer-Verlag, N Y.

Pimentel, D. 1986b. Using genetic engineering for biological control: reducing ecological risks. Pp. 129-140 *in* Halvorson, H. O., D. Pramer, and M. Rogul (eds.), Engineered organisms in the environment: scientific issues. American Society for Microbiology, Washington, D. C.

Regal, P. J. 1986. Models of genetically engineered organisms and their ecological impact. Pp. 111-129 *in* Mooney, H. A., and J. A. Drake, Ecology of biological invasions of North America and Hawaii. Ecological Studies 58. Springer-Verlag, N Y.

Rifkin, J., and P. Nicanor. 1983. Algeny. Viking, N.Y.

Sayler, G., and G. Stacey. 1986. Methods for evaluation of microorganism properties. Pp. 35-55 *in* Fiksel, J., and

V. T. Covello (eds.), Biotechnology risk assessment: issues and methods for environmental introductions. Pergamon, N Y.
Schoener, T. W. 1983. Field experiments on interspecific competition. American Naturalist 122:240-285.
Sharples, F. E. 1983. Spread of organisms with novel genotypes: thoughts from an ecological perspective. Recombinant DNA Technical Bulletin 6:43-56.
Sharples, F. E. 1986. Evaluating the effects of introducing novel organisms into the environment. United Nations Environment Programme, Nairobi Kenya. [See Sharples 1987 for availability.]
Sharples, F. E. 1987. Regulation of products from biotechnology. Science 235:1329-1332.
Simberloff, D. S. 1980. A succession of paradigms in ecology: essentialism to materialism and probabilism. Synthese 42:3-39.
Simberloff, D. S., and R. K. Colwell. 1984. Release of engineered organisms: a call for ecological and evolutionary assessment of risks. Genetic Engineering News 4(7):4.
Sousa, W. P. 1984. The role of disturbance in natural communities. Annual Review of Ecology and Systematics 15:353-391.
Strong, D., J. H. Lawton, and T. R. E. Southwood. 1984. Insects on plants: community patterns and mechanisms. Blackwell Scientific Publications, Oxford.
Williams, M. C. 1980. Purposefully introduced plants that have become noxious or poisonous weeds. Weed Science 28:300-305.
Young, J. A., R. A. Evans, R. E. Eckert, and G. L. Kay. 1987. Cheatgrass. Rangelands (in press).
Zimmerman, B. K. 1984. Biofuture: confronting the genetic era. Plenum, N.Y.

Plasmid Transfer in a Freshwater Environment

M. J. Day, M. J. Bale and J. C. Fry
Department of Applied Biology,
University of Wales Institute of Science and Technology,
PO Box 13,
Cardiff CF1 3XF.
Wales.

1. Introduction

As the potential uses for genetically manipulated micro-organisms widen the demand for their exploitation will increase. This will inevitably result in their deliberate or accidental release, an occurrance which has led ecologists to predict environmental disaster [13]. Others believe that the chance of disaster is remote because of experience with non-genetically engineered organisms [10].

The assessment of risk, due to release of any organism into an environment, involves an evaluation of the organism's ability to survive, reproduce and become established or transported into another environment. Any invasive and pathogenic properties must also be evaluated. Additional criteria operate with genetically manipulated organisms. The persistence of the manipulated organism's DNA is one element. Intimately linked to this is the possibility of gene transfer from the manipulated micro-organism into the indigeneous bacterial populations.

This is one of the areas which was nominated for further research in a book entitled 'Biotechnology Risk Assessment' [12] which stated that an *investigation of plasmid transfer and other ecological aspects of the fate of genetic material in the environment was essential.*

The ecological complexity of the natural environment coupled with our poor understanding of it makes it unlikely that we

NATO ASI Series, Vol. G18
Safety Assurance for Environmental Introductions
of Genetically-Engineered Organisms
Edited by J. Fiksel and V. T. Covello

will be able to accurately predict the fate and persistence of a manipulated micro-organism or its genes in the near future. However once the basic rates of bacterial persistence and gene transfer are established a more realistic assessment can be made.

In this chapter a method for establishing a base line rate for natural, plasmid-mediated gene-exchange in a freshwater system is presented. It has been developed using indigenous (natural) plasmids and laboratory bacteria, but could be adapted to examine transfer from genetically engineered micro-organisms. With some modifications, to suit particular test sites, it could be used to examine gene transfer in any environment.

2. Modes of and Limits to Gene Transfer

There are three mechanisms of gene transfer, transformation, transduction and conjugation [30]. Transformation occurs naturally in a few genera [33] and molecular DNA, released by a cell during growth or by lysis, being taken up by a physiologically competent cells. Transduction, or phage mediated gene transfer, which involves the packaging of host genes into the phage head has limitations. This is because we believe most phage to have a narrow host range, although this still has to be tested [30]. The frequency of bacteria, capable of transferring genes in either of these ways, within a community, would be important to any risk assessment. The third transfer mechanism is conjugation, a process of gene exchange mediated by a plasmid and involving cell to cell contact. Other genes (from non-conjugative plasmids or from the chromosome) can be transferred by the conjugative plasmid. In some cases this transfer is confined to related bacteria, but plasmids with extremely broad host ranges are known. Plasmids can also be transferred by transduction and transformation [31].

It is probably that all bacteria are capable of genetic

exchange by at least one of these mechanisms, indeed as our knowledge increases it is obvious that these are the ways in which the genetic diversity, of microbial populations is maintained.

There is a mechanism termed restriction which operates to reduce these processes of genetic communication [1]. It can adversely influence the success of transfer between laboratory bacteria and can significantly effect even quite closely related strains. Many bacteria have such a host mediated restriction/modification system which surveys incoming DNA. If the DNA is host labelled (modified) then it evades digestion by the host specified restriction endonuclease system. Other limitations are imposed by the host ranges of the transfer mechanisms and the ability of the new host to stabilise either the gene by recombination or plasmid by replication.

Viable and closely associated donor and recipient cells are essential for conjugation, but not for transduction or transformation. In these two cases the fact that the donor and recipient cells can be separated in both time and space is an important criteria for consideration in risk assessment. Finally the physical aspects of the environment, for example temperature, pH, salinity, nutrient status, etc. may each influence the potential for gene transfer.

3. Transfer in Natural Systems

Attempts have been made to show that transfer occurs in a variety of natural habitats. Many studies have demonstrated gene transfer in the mammalian gut and most have used plasmid encoded antibiotic resistances in *Escherichia coli*. Transfer is most frequent during starvation, during antibiotic treatment, in very young and in gnotobiotic animals, but is rarely observed in the normal gut populated by an undisturbed microflora [14]. Plasmid transfer also occurs on plant surfaces, between *Klebsiella spp.* on whole radish plants and between phytopathogenic *Pseudomonas spp.* on bean leaves [22, 26, 34]. In aquatic environments gene transfer has been demonstrated in containers but not when the bacteria are unenclosed. Transfer of plasmid-borne antibiotic resistance has been shown in membrane diffusion chambers suspended in waste water treatment plants [2, 24] and dialysis sacs in freshwater [17, 18]. These experiments were all done with pure cultures of enterobacteria and in only one instance were the competing indigenous microflora present [18]. Unenclosed experiments are important if plasmid or gene transfer rates in aquatic habitats are to be calculated.

The results of work presented here can be used to identify factors which maybe important in the processes of risk assessment. We chose the epilithon (the slimy community on the surface of submerged stones) because it has a dense population of closely spaced bacteria [23] which are fixed and hence more likely to transfer plasmids by conjugation than other communities of aquatic bacteria. *Pseudomonas spp.* were used because these are very commonly occurring aquatic bacteria [19, 20, 29] which often contain plasmids [11].

4.1. Conjugation Methodology

The full description of the media, protocols and strains used have been published [4] but a brief description will be given here. In laboratory matings aliquots of an overnight culture of donor (0.2 ml) and recipient (1.8 ml) were mixed and filtered through a 0.22 μm cellulose acetate filter. These were placed face up on agar plates and incubated for 24 h at the required temperature and donors, recipients and transconjugants were enumerated on selective media.

4.2. Isolation of the Natural Plasmid pQM1

The epilithic bacteria were removed from flat stones, taken

from the River Taff, an organically polluted river in South Wales [28]. The bacteria was mixed with a streptomycin resistant recipient, *P. aeruginosa* PAO2002R, and deposited onto a membrane as described elsewhere [4]. Following incubation selection was made for mercury resistant PAO2002R. Several mercury resistant recipients were identified. These transconjugants were screened for plasmids by gel electrophoresis and for the ability to transfer mercury resistance into a rifampicin resistant recipient, PU21. One mercury-resistance plasmid, pQM1, was isolated by this method and used in subsequent studies. It is a large (255 kb), narrow host range *Pseudomonas* plasmid belonging to the *Inc*P-13 group and produces thick flexible pili typical of plasmids that transfer well in both liquid and solid phases [9].

4.3. Microcosm Experiments

Studies of plasmid transfer with enclosures in aquatic habitats have found transfer frequencies to be lower than those found in laboratory experiments [2, 17, 24]. Therefore, preliminary laboratory microcosm experiments simulating natural conditions were used to ensure that the methodology would work *in situ*.

Aliquots of PAO2002R (pQM1) and PU21 were deposited on membrane filters which are placed face down on a sterile scrubbed stone, secured with a rubber band and placed either in PCA broth, minimal salts medium or non-sterile river water [4]. The beaker was sealed and incubated at 6°C or 20°C for 24 h. The bacteria on the stone, filter and in the liquid phase were harvested and enumerated for donors, recipients, and transconjugants on media containing streptomycin, rifampicin and rifampicin with mercury, respectively, as selective agents.

4.4. *In situ* Experiments

In subsequent transfer experiments the bacteria were deposited on sterile membrane filters and incubated face down on sterile scrubbed stones either in a canal fed by the River Taff or in the river itself. The mean number of donors on the membrane filters was 5.3 x 10^7 cm^{-2} which was similar to the total bacterial count of the natural epilithon (direct count = 1.6 x 10^8 bacteria cm^{-2}). The chance of detecting transfer was maximised by using a recipient to donor ratio of 12:1.

4.5. Experiments on Unscrubbed Stones

The effect of the natural epilithon on plasmid transfer was investigated by matings on unscrubbed stones in laboratory microcosm overlain with non-sterile river water or *in situ*. To investigate the physical effects of the epilithon on conjugation stones were pasteurized by heat at 65°C for 10 mins or autoclaved at 121°C for 15 min. All matings were done as described elsewhere [4].

4.6. Experiments Using Integrated Bacteria

Laboratory donors and recipients were incubated in contact with the epilithon until they became integrated into the microbial community. Mating experiments were then done with these bacteria. Aliquots of laboratory donors, (PAO2002R, pQM1) and recipients (PU21) were deposited onto separate filters which were then placed face down on stones freshly collected from the river. Incubation for 24 h *in situ* was sufficient to incorporate the bacteria into the epilithon. The filters were removed and the stones were rinsed in river water. Matings were begun by placing the two stones in contact so that the area where the filters had been were touching. The stones were incubated *in situ* for 24 h. The stones were removed, separated and sampled. The bacteria were enumerated as described above. Full details of this and related methods are given elsewhere [7].

5. Results and Discussion

5.1. Microcosm Experiments

It was clear from the microcosm experiments (Table 1) that the transfer frequency of pQM1 was significantly higher ($P < 0.001$) in the complex medium than when the stones were incubated in river water. The results for the mineral salts medium were intermediate. Contact with the stones had no effect on transfer as similar matings on agar were not significantly different from the microcosm results. With all the suspending media, transfer at 6°C was at least a thousand-fold less than transfer at 20°C. The direct contact between the membrane filter and the sterile stone clearly resulted in adherence of organisms to the stone surface, as donors, recipients and transconjugants were present on the stones after incubation (Table 1). *P. aeruginosa* is motile and had therefore moved from the filters and stones into the liquid phase in all the incubations. After incubation at 20°C significantly more donor and recipient cells were found in the liquid phase than were originaly put onto the membrane filters hence growth had occurred. However, at 6°C the results from counts of donors and recipients indicated that cells had become non-recoverable during incubation. The transfer frequency in the liquid phase was very low at 6°C (Table 1), despite there being more donors and recipients present than on the filter or stone. This implied that movement of the bacteria into the liquid medium occurred before mating and so most plasmid transfer had occurred on the stone and filter surfaces.

5.2. *In situ* Experiments

Transfer experiments were carried out *in situ* that were comparable to the microcosm experiments. Stones with attached membrane filters, inoculated with donors and recipients, were put into open-topped beakers submerged in the Taff Feeder

Table 1: The effect of temperature and suspending medium on transfer frequencies of pQM1 between *Pseudomonas aeruginosa* strains on sterile scrubbed stones incubated in laboratory microcosms.

Incubation temperature (°C)	Suspending medium	Number of replicate experiments performed	Transfer frequency per recipient on: filter	stone	liquid phase
20	PCA broth	2	3.8x10^{-2}	3.3x10^{-1}†	1.9x10^{1}†
	Minimal salts	2	9.3x10^{-3}	1.3x10^{-2}	1.1x10^{-2}
	R. Taff water	1	1.9x10^{-3}	7.5x10^{-3}	1.1x10^{-2}
6	PCA broth	3	3.4x10^{-5}	1.0x10^{-5}	4.4x10^{-6}
	Minimal salts	1	1.0x10^{-5}	5.7x10^{-6}	<3x10^{-6}+
	R. Taff water	2	1.6x10^{-6}*	1.1x10^{-6}*	<5x10^{-9}+

Mating experiments using PAO2002R (pQM1) donors and PU21 recipients were done on filters on PCA plates or on scrubbed stones in three different suspending media in simple micrcosms for 24 h. Control matings on PCA agar plates gave pQM1 transfer frequencies of 2.8 x 10^{-1} at 20°C and 1.6 x 10^{-5} at 6°C [4].

PCA = Standard plate count agar.

Average coefficient of variation between replicate experiments = 8.7%

Minimum significant difference ($\log_{10}x$) = 0.56

* significantly lower than all other frequencies

† significantly higher than all other frequencies

+ Transfer not detected.

Canal. This canal receives its water from the River Taff, but has a much lower flow rate than the main river and so is ideal for experimental use. Transfer of pQM1 occurred in these beakers at environmental temperatures, between March and November 1986 and a graph of conjugation frequency and river temperature over the 7 month period is shown in Figure 1 [6].

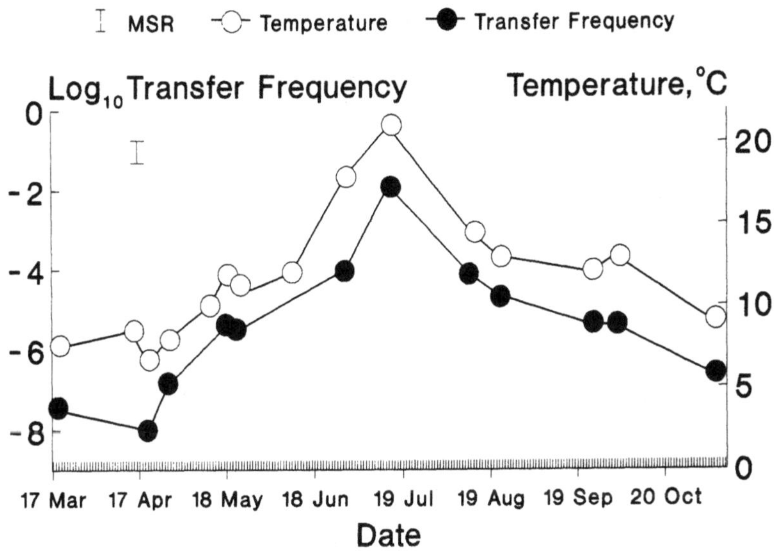

Figure 1 The frequency of pQM1 transfer from PAO2002R(pQM1) to PU21 on scrubbed stones incubated in the Taff Feeder Canal. The river temperature over the same period is also shown.

As can be seen the temperature rose from 6-7°C in March and April to a maximum of 21°C in July and then declined to 9°C by early November. The curve of transfer frequency followed the temperature closely high frequencies were associated with the highest temperatures, as was expected from the results of the microcosm experiments. Linear regression of the logarithm of transfer frequency against temperature showed a highly significant regression (Figure 2). The regression line explains 95% of the variation in transfer frequencies suggesting the primary importance of temperature. The slope of this line is 0.38, indicating that a 2.6°C increase in temperature leads to a 10-fold increase in transfer frequency. Temperature variations of greater than 2.5°C occurred over relatively short periods in the River Taff, for instance, between late April and late May the temperature rose from 6°C to 11°C and between July and August the temperature was between

15 °C and 21 °C (Figure 1). Comparison of the *in situ* frequencies at 21 °C (1.5 x 10^{-2} per recipient) with the frequencies found in microcosms using river water as suspending medium at 20 °C (3.7 x 10^{-3}) showed that the *in situ* transfer frequencies were

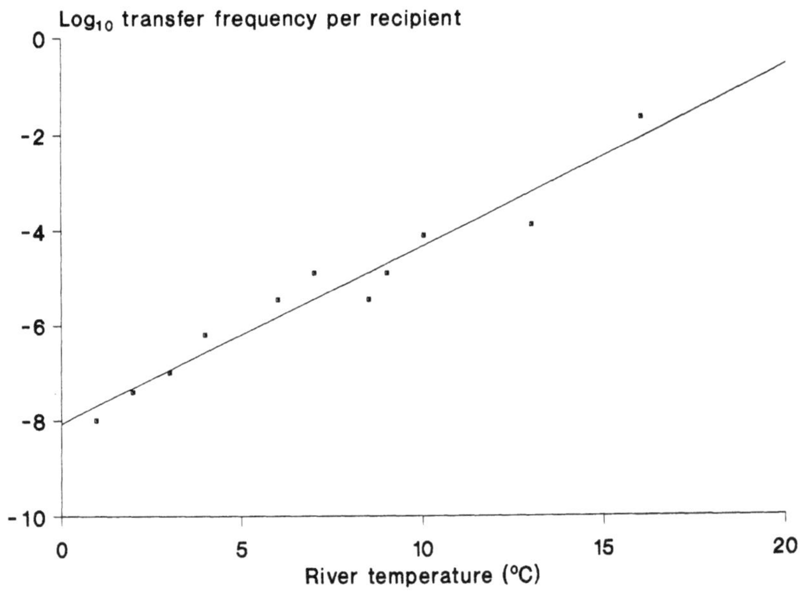

Figure 2 regression line of the $\log_{10}$ transfer frequencies against river temperature for the *in situ* transfer of pQM1 from PAO2002R(pQM1) to PU21.

significantly higher. Control experiments were done to ensure no transfer occurred during transit, the frequencies were 100 times lower than the corresponding 24 h matings [4].

The effect of temperature on the transfer frequency of several plasmids has been reported by others. The *E. coli* plasmid R1*drd*-19 transferred at 17 °C over 24 h but not at 15 °C in 48 h [32]. A mercury resistance plasmid from a marine pseudomonad transferrred to *E. coli* at 25 °C but not at 16 °C [15] and the plasmid pWK1 transferred between *Erwinia herbicolor* strains at 12.5 °C [21, 30]. However, two reports have been made of transfer at lower temperatures similar to those we have used. One reported transfer at 10.6 °C, between a range of enterobacteria [24] and another [35] showed a graph with no experimental details of IncFII and *Inc*Iα plasmids transferring at 4°C. It therefore seems probable that transfer of plasmids between indigeneous aquatic bacteria can occur at environmentally relevant temperatures in the epilithon. However, transfer frequencies will probably be lower in the colder waters of winter than in the warmer summer months.

5.3. Experiments on Unscrubbed Stones

Further experiments were done using stones with the intact epilithon to see if conjugation would still occur in the presence of the full, competing, natural microflora (Table 2). Although transfer was not detected at 7°C it clearly occurred at 11 and 12°C in both laboratory and *in situ* experiments. The transfer frequencies were significantly lower than those obtained at similar temperatures using sterile scrubbed stones (Table 1). This was presumably due to competition from the natural microflora. When the natural epilithon was killed by heating, the transfer frequencies of pQM1 were significantly greater than those on untreated stones (Table 2) and were almost all significantly higher than those from sterile scrubbed stones (Table 1, Figure 2). The enhancement of transfer on pasteurised stones was greater than that on autoclaved stones and this transfer frequency was only about 28 times less than the value obtained in a control laboratory experiment at 12°C on PCA agar (9.2×10^{-3} per recipient) and similar to what we would expect in PCA broth at a similar temperature (Table 1). Consequently, it is probable that heating enhances transfer by killing the natural competing microflora and by releasing nutrients from the polysaccharide matrix known to be present in the epilithon [23]. Thus in addition to temperature, nutrient status appears to be a major factor influencing the rate of pQM1 transfer. These observations are supported by Beringer [8] who examined RP4 plasmid transfer between two *Rhizobium leguminosarum* strains. He showed that reductions in the concentrations of nutrient broth or glycine (50% - 0%) in the mating media lowered the frequency of transfer between 100 to 300-fold.

Table 2: Transfer frequencies of pQM1 between *Pseudomonas aeruginosa* strains on unscrubbed stones from the R. Taff incubated in laboratory microcosms and the Taff Feeder Canal.

Experimental site and experiment number	Treatment	Temperature (°C)	Frequency per recipient on Filter	Stone
Laboratory				
1	Nil	7	$<5\times10^{-9}$*	$<5\times10^{-9}$*
2	Nil	12	4.9×10^{-6}	5.0×10^{-7}
3	Nil	11	1.7×10^{-6}	4.9×10^{-7}
1	Autoclaved	7	2.2×10^{-6}	2.5×10^{-6}
2	Pasteurised	12	3.3×10^{-4}	7.8×10^{-4}
In situ				
3	Nil	11	6.3×10^{-7}	4.0×10^{-7}
3	Autoclaved	11	3.1×10^{-6}	7.2×10^{-7}
3	Pastuerised	11	7.8×10^{-6}	3.1×10^{-5}

The matings using PAO2002R (pQM1) donors and PU21 recipients were similar to those using sterile scrubbed stones. The stones were either untreated (Nil), heated at 65°C for 10 min (Pasteurized) or at 121°C for 15 min (Autoclaved). The experiment number indicates the results that were obtained on the same date.

* transfer not detected.
Minimum significant difference ($\log_{10}x$) = 0.41; Average coefficient of variation between replicate experiments = 2.3%

5.4. Integration of Bacteria into the Epilithon

The results above showed that donors or recipients could survive in the epilithon for limited periods and so work was done to investigate transfer between donors and recipients growing as an integrated part of the epilithon. This could represent a more natural system for studying plasmid transfer[7].

The results from two mating experiments *in situ* are shown in Figure 3. A control filter mating on a sterile stone is shown for comparison. The frequencies between the two experiments varied by a factor of 10^5 whilst the temperature only varied by 1.5°C. This variation appeared to be due to the donor:recipient ratio, because the transfer frequencies were highest when the ratio was highest (672:1; 2.2×10^{-1} per recipient). The reason for the variation in donor to recipient ratios and transfer frequencies was the migration of donors and recipients between their respective stones. Approximately 0.7% to 2% of the integrated donors and recipients migrated giving ratios from 672:1 to 1:170. The epilithon consists of microcolonies of daughter cells [23] which may occasionally come into contact with migrant individual bacteria leading to ratios of 20:1 to 1:20, which is near optimum for pQM1 transfer.

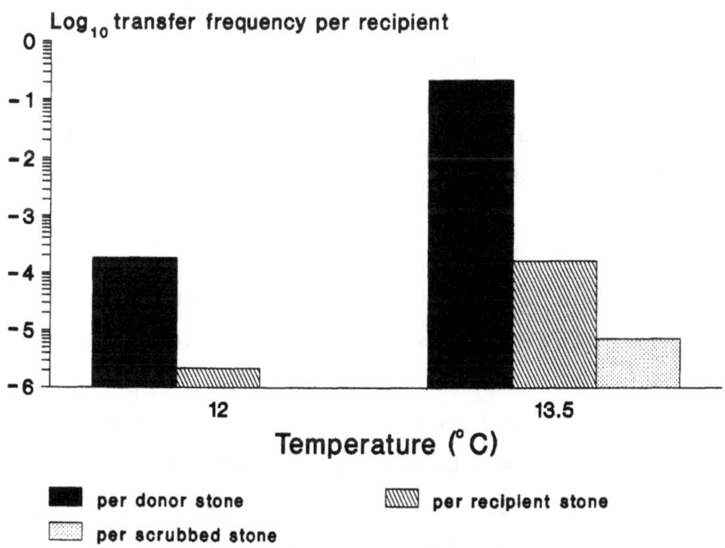

Figure 3 Histogram showing the transfer frequencies of pQM1 between PAO2002R(pQM1) and PU21 incorporated into the epilithon on separate stones. the frequencies are for the stone containing donor and recipient as well as sterile scrubbed stone incubated *in situ*. The donor to recipient ratios were 2.6:1, 1:166, 672:1, 1:17 and 1:27 respectively.

The densities of laboratory bacteria on the stones in these experiments were low, between 1.1 x $10^3 cm^{-2}$ and 1.6 x $10^6 cm^{-2}$, or from 0.0005% to 0.5% of the epilithic bacterial density. Thus transfer can occur at very low bacterial densities. Other studies of *in situ* transfer found high bacterial densities were needed to detect transfer, usually higher than or comparable with the indigenous bacterial densities [2, 17, 24, 26, 27].

To summarize; it proved possible to use a natural plasmid isolated from the epilithon to study transfer between laboratory strains in an aquatic habitat. Experiments showed that temperature, but also nutrient status, were important factors in affecting conjugation. Competition with natural bacteria did reduce transfer but it was still possible to detect transfer when the laboratory donor and recipients were growing and integrated into the epilithon. The epilithon proved to be a useful natural system which could be easily manipulated and contained large numbers of active bacteria and plasmids. Our studies of epilithic bacteria clearly show large numbers of plasmids and the presence of actively conjugative plasmids [6] suggesting high stability or persistance and little or no disadvantage to plasmid containg hosts. These results complicate the assessment of naturally occuring gene transfer because they prompt several further questions.

Firstly, how common are conjugative plasmids with temperature optima at other temperatures? Secondly, how common are broad host range plasmids and to what extent does the host and recipient affect plasmid transfer? Thirdly, how efficient are conjugative plasmids at mobilising small plasmids? Further work on mobilization is needed particularly in light of the results of Gealt and coworkers [16, 25, 27].

6. Conclusions

The overall conclusion from this work is that plasmid transfer occurs between bacteria in the river epilithon. This work shows that transfer between organisms actually incorporated into the epilithon can be detected. This is the only method currently available which can be used as a model system for studying conjugation in aquatic habitats. If pQM1 and the pseudomonad host strains used are typical of indigenous epilithic bacteria, then these plasmids will spread rapidly throughout riverine bacterial populations. The results suggest that the overall effect of temperature is to enhance plasmid transfer throughout rivers in summer and depress it during winter. Although the River Taff is organically polluted the sites investigated were not near oganically rich effluents. Consequently it is also probable that plasmid transfer will be greatest where available organic substrate concentrations are highest, such as directly below sites of organic enrichment.

7. References

[1] Arber, W. & S. Linn.(1969) DNA modification and restriction. *Ann. Rev. Biochem.* *37*:467-500.

[2] Altherr, M. R. & K. L. Kasweck. (1982) *In situ* studies with membrane diffusion chambers of antibiotic resistance transfer in *Escherichia coli*. *Appl Environ. Microbiol.* *44*: 838-843.

[3] Bale, M. J. (1987) Plasmid transfer between bacteria in natural epilithon. Ph.D. thesis, U.W.I.S.T., Cardiff.

[4] Bale, M. J., J. C. Fry and M. J. Day (1987a) Plasmid transfer between strains of *Pseudomonas aeruginosa* on membrane filters attached to river stones. *J. Gen. Microbiol.* *133*: 3099-3107..

[5] Bale, M. J., Rochelle, P. A., Day, M. J. & J. C. Fry (1987b) Comparison of three media for studying plasmids in epilithic bacteria with particular reference to conjugative mercury resistance plasmids. Submitted to *J. Appl. Bacteriol.*

[6] Bale, M. J., J. C. Fry and M. J. Day (1987b) Transfer and occurence of large mercury-resistance plasmids in river epilithon. Submitted to *Appl. Environ. Microbiol.*

[7] Bale, M.J., M.J. Day, & J.C. Fry (1987d) A novel method for studying plasmid transfer in undisturbed river epilithon. Submitted to *Letters in Applied Microbiology.*

[8] Beringer, J. E. (1974) R factor transfer in *Rhizobium leguminosarum*. *J. Gen. Microbiol., 84*: 188-198.

[9] Bradley, D. E. (1983) Specification of the conjugative pili and surface mating systems of *Pseudomonas* plasmids. *J. Gen. Microbiol.* *129*: 2545-2556.

[10] Brill, W. J. (1985) Softer concerns regarding genetically engineered plants and microorganisms to benefit agriculture. *Science 227:* 381-384.

[11] Burton, N. F., Day, M. J. & A.T. Bull. (1982) Distribution of bacterial plasmids in clean and polluted sites in a South Wales River. *Appl. Environ. Microbiol. 44*: 1026-1029.
[12] Fiksel, J. & V. Covello (1986) Biotechnology and Risk Assessment: Issues and Methods for Environmental Introduction. Fiksel, J. and V. Covello, (Eds.), Pergamon Press.
[13] Forcella, F. (1984) Computary : Ecological biotechnology *Bull. Ecol. Soc. America 64*: 434-436.
[14] Freter, R. (1984) Factors affecting conjugal plasmids transfer in natural bacterial communities. In M. J. Klug and C. A. Reddy (Eds.), *Current Perspectives in Microbial Ecology.* Washington D.C. American Society for Microbiology, pp 105-114
[15] Gauthier, M. T., F. Cauvin, & J-P. Breittmayer, (1985) Influence of salts and temperature on the transfer of heavy metal resistance from a marine pseudomonad to *Escherichia coli. Appl. Environ. Microbiol. 50*: 38-40.
[16] Gealt, M. A., M. D. Chai, D., K. B. Alpert, J. C. & Bover, (1985) Transfer of plasmid pBR322 and pBR325 in wastewater from laboratory strains of *Escherichia coli* to bacteria indigenous to the waste disposal system. *Appl. Environ. Microbiol.* 49: 836-841
[17] Gowland, P. C. & J. H. Slater. (1984) Transfer and stability of drug resistance plasmids in *Escherichia coli* K12. *Microbial Ecol. 10*: 1-13.
[18] Grabow, W. O. K., O. W. Prozesky, & J. S. Burger. (1975) Behaviour in a river and dam of coliform bacteria with transferable or non-transferable drug resistance. *Water Res. 9*: 777-782.
[19] Holder-Franklin, M. A., M. Franklin, P. Cashion, C. Cormier, & L. Wuest. (1978) Population shifts in heterotrophic bacteria in a tributary of the Saint John River as measured by taxometrics. In M. W. Loutit and J. A. R. Miles (Eds.), *Microbial Ecology* Berlin: Springer Verlag, pp44-50.
[20] Jones, J. G., S. Gardener, B. M. Simon, & R. W. Pickup. (1986) Antibiotic resistant bacteria in Windermere and two remote upland tarns in the English Lake District. *J. Appl. Bacteriol. 60*: 443-453.
[21] Kelly, W. J. & D. C. Reanney. (1984) Mercury resistance among soil bacteria: ecology and transferability of genes encoding resistance. *Soil Biology and Biochem. 16*: 1-8.
[22] Lacey, G. H. & J. V. Leary. (1975) Transfer of antibiotic resistance plasmid RP1 into *Pseudomonas glycinea* and *Pseudomonas phaseolicola in vitro* and *in planta. J. Gen. Microbiol. 88*: 49-57.
[23] Lock, M. A., R. P. Wallace, J. W. Costerton, R. M. Ventullo, & S.E. Charlton. (1984) River epilithon: toward a structural functional model. *Oikos, 42*: 10-22.
[24] Mach, P. A. & D. J. Grimes. (1982) R-plasmid transfer in a waste water treatment plant. *Appl. Environ. Microbiol. 44*: 1395-1403.
[25] Macherson, P. & M. A. Gealt, (1986) Isolation of indigeneous wastewater bacterial strains capable of mobilising plasmid pBR325. *Appl. Environ. Microbiol. 51*:904-909.
[26] Manceau, C., L. Gardan, & M. Devaux, (1986) Dynamics of RP4 plasmid transfer between *Xanthamonas campestris* pv *corylina*

and *Erwinia herbicola* in hazelnut tissue *in planta. Can. J. Microbiol. 32*: 835-841.
[27] Mancini, P., S. Fertals, D. Nave, & M. A. Gealt, (1987) Mobilisation of plasmid pHSV106 from *Escherichia coli* HB101 in a laboratory-scale waste treatment facility. *Appl. Environ. Microbiol. 53*: 665-671.
[28] Mawle, G. W. Winstone, & M. P. Brooker. (1985) Salmon and sea trout in the Taff - past, present and future. *Nature in Wales New Series 4 (102)*: 36-45.
[29] Nuttall, D. (1982) The populations, characterisation and activity of suspended bacteria in the Welsh River Dee. *J. Appl. Bacteriol. 53*: 49-59.
[30] Reanney, D. C., W. P. Roberts & W. J. Kelly (1982) Genetic interactions among microbial communities. In A. T. Bull and J. H. Slater (Eds.), *Microbial Interactions and Communities*, Academic Press, London pp 289-322.
[31] Saye, D. J., G. S. Ogunseitan, G. S. Sayler, & R. V. Miller (1987) Potential for transduction of plasmids in natural freshwater environment: effect of plasmid donor concentration and a natural microbial community on transduction in *Pseudomonas aeruginosa. Appl. Environ. Microbiol 53*:987-995.
[32] Singleton, P. & A. E. Anson. (1981) Conjugal transfer of R-plasmid R1*drd*-19 in *Escherichia coli* below 22°C. *Appl. Environ. Microbiol. 42*: 789-791.
[33] Stewart, G. J. & C. A. Carlson (1986) The biology of natural transformation. *Ann. Rev. Microbiol. 40*: 211-235.
[34] Talbot, H. W., D. K. Yamamoto, M. W. Smith, & R. J. Seidler. (1980). Antibiotic resistance and its transfer among clinical and non-clinical *Klebsiella* strains in botanical environments. *Appl. Environ. Microbiol. 39*: 97-104.
[35] Timoney, J. F. (1981) R-plasmids in pathogenic Enterobacteriaceae from calves. In S. B. Levy, R. C. Clowes and E. L. Koenig (Eds.), *Molecular Biology, Pathogenicity and Ecology of Bacterial Plasmids.* London: Plenum Press, pp

547-555.

L-FORM BACTERIA AND SOMATIC ASSOCIATIONS WITH EUCARYOTES

A.M. PATON
SCHOOL OF AGRICULTURE
UNIVERSITY OF ABERDEEN
ABERDEEN, AB9 1UD
SCOTLAND, UK.

Within the context of a workshop concerned with aspects of the environmental releases of genetically engineered organisms it is appropriate to consider the implications of an alternative means of "engineering" whereby desired expressions are obtained by creating stable somatic associations and hence avoiding actual genetic manipulation.

An outline of L-form bacteriology and the phenomenon of association with eucaryotes.

L-form bacteria have been recognised since 1935 and have been the subject of mainly academic study up to the present time (e.g. Guze 1968, Hayflick 1969 & Madoff 1986). They are, essentially, ordinary bacteria which have been enabled to grow without their cell wall components. This change can be brought about by culturing in the presence of inducing agents, which inhibit or otherwise interfere with the production of cell-walls, such as penicillin or lysozyme. The resulting cells can be provided with mechanical stability by ensuring an adequate osmotic status in the medium. L-form bacteria can then be grown in liquid or on solid media where they generally demonstrate a complicated life cycle not normally observed with bacteria.

The cells have a very varied morphology both in size and internal structure. The sizes may range from twenty times that of the parent organism to viable units capable of passing through 0.22µ pore membranes. Generally these L-form cells will regain their cell walls when inducing agents are omitted and are therefore considered unstable. However, by careful selection stable cultures may be obtained and these can be continuously cultivated in the absence of inducing agents.

When either stable or unstable cultures of L-form bacteria are brought into intimate contact with plant or lower eucaryotic cells entry is achieved and a permanent association is created, the L-forms maintaining their life-cycle within the host cells without any apparent deteriorating

NATO ASI Series, Vol. G18
Safety Assurance for Environmental Introductions
of Genetically-Engineered Organisms
Edited by J. Fiksel and V.T. Covello

effect on either partner. By this means somatic associations of both Gram positive and Gram negative bacteria have been formed with plant tissue cultures, whole plants, and with fungi such as yeasts. These associations are found to provide the host cells or tissues with biochemical expressions related to those of the enclosed bacterial L-forms (Aloysius & Paton 1984). Thus an unlimited number of novel associations can be constituted using appropriate bacteria without the involvement of gene transfer (Aloysius & Paton 1984, Paton 1987).

The implications and potential applications of the technique

There now exists an ability to form an infinite number of somatic associations of normal bacteria with plants, fungi and, possibly, with other eucaryotes. By reason of metabolic interaction it is unlikely that all desired expressions of the L-forms will be revealed when an association is formed but experience has shown that the vast choice of partnerships increases the opportunities for success. The work so far carried out has involved mainly non-engineered bacteria but the applications of the technique can, obviously, be greatly extended by the use of engineered bacteria as L-forms. A further extension of the applications can be envisaged by the incorporation of multiple strains of L-form bacteria. The nature and outcome of the resulting complicated interactions can only be, presently, a matter of speculation. Although the associations already constructed have practical value in their own right a further outcome of this methodology is the use of L-forms within eucaryotes as vectors for genetic engineering purposes.

The practical applications envisaged are many and widely varied. Among those which are possible within the near future are modifications in plant growth and the production of novel metabolites in plants and fungi. Mycorrhizal manipulation with selected associations and biological disease and pest control would be likely to receive a similar priority.

The relationship with genetic engineering in respect of risk assessment

Although risk assessment is presently directed towards the control, where desirable, of organisms which have been genetically altered or modified, the implications of the introduction into the environment of somatically associated organisms have not been seriously considered by advisory or legislative agencies. The involvement of genetically engineered bacteria in the described associations would undoubtedly come under the present parameters of risk assessment.

The endocytic relationships of L-forms originating from normal or "natural" bacteria with higher life forms do, however, provide novel ecological niches for these bacteria allowing large populations to exist in entirely new environments. The L-form bacteria are essentially very vulnerable to both the physical and chemical influences of general environment.

Although the somewhat related structures of the mycoplasmas do exist in plant and animal systems there is no evidence for more than a transient free-living state in nature for the L-forms. No natural associations of the type described above have so far been recognised in plant cells. According to present information the spread of the L-form bacteria in host tissue depends on the formation of the "particles" or "granules" within the bacterial cell as part of its life-cycle. Following general dissemination such particles would only initiate a further association in the unlikely event of surviving and entering the acceptable environment of a new host cell. No instance of this event occurring has been observed. L-form bacteria which have been passed intracellularly through a plant system have maintained thereafter a stability and have not reverted under _in vitro_ or _in vivo_ conditions. This suggests that L-form bacteria making up, say, a plant association will not, on release, be capable of initiating a population of the parent bacteria. It would appear that somatic associations using originally non-hazardous bacteria converted to L-forms will constitute a minimal risk but further work is necessary to increase the degree of certainty.

References

Aloysius, S.K.D. & Paton, A.M. 1984. Artificially induced symbiotic associations of L-form bacteria and plants. _Journal of Applied Bacteriology_ 56, 465-477.

Guse, L.B. 1968 _Microbial Protoplasts, Spheroplasts and L-forms._ Baltimore: Williams & Wilkins.

Hayflick, L. 1969 The Mycoplasmatales and the L-phase of bacteria. New York: Appleton-Century Crofts.

Madoff, S. 1986 The Bacterial L-forms. New York & Basel: Dekker.

Paton, A.M. 1987 L-forms: Evolution or revolution? Journal of Applied Bacteriology 63, 365-371.

Deliberate Release of Genetically Engineered Organisms in the Environment

Daphne Kamely
Office of Research and Development (RD-689)
U.S. Environmental Protection Agency
Washington, DC 20460[1]

Introduction

When recombinant DNA technology was first developed, it was limited to basic research laboratories performing primarily biomedical research. As most biomedical research in the United States is funded by the National Institutes of Health (NIH), the 1976 NIH Guidelines for Research Involving Recombinant DNA Molecules [1] adequately covered all Government funded research. Other research institutions and most industries performing recombinant DNA research voluntarily agreed to abide by the NIH Guidelines. This voluntary control continued to work well without established regulations until recombinant DNA research moved from the public to the private sector, where it was expanded to include other biotechnical advances such as hyubridoma research; the large-scale production of new drugs, hormones, chemicals, enzymes, diagnostic and detection devices; as well as novel process technology. Scientists were no longer experimenting with small 10-liter quantities; large scale fermentation processes were required. In agriculture and in industrial areas, research started to move from the laboratory into the open field. The NIH Guidelines could no longer cover this rapid growth. The intervention of the regulatory agencies became necessary as products moved from the laboratory to the market place.

---

[1]Current Address: Scientific Advisor for Biotechnology, U.S. Army CRDEC, SMCCR-TDB, Aberdeen Proving Ground, Maryland 21010.

Disclaimer: These opinions are the Author's own and not necessarily those of the U.S. Environmental Protection Agency. No official support or endorsement t by EPA or any other agency of the U.S. Government is intended or should be inferred.

NATO ASI Series, Vol. G18
Safety Assurance for Environmental Introductions of Genetically-Engineered Organisms
Edited by J. Fiksel and V. T. Covello

In this paper, I focus on the mandate of regulatory agencies that have jurisdiction to regulate biotechnology research and development in the United States. In particular, I address the deliberate release of recombinant DNA molecules and novel organisms into the environment in the context of the risk assessment and regulatory programs in the U.S. Environmental Protection Agency (EPA), the primary Federal Agency responsible for the regulation of biotechnology products and processes released into environment and the assurance of their safety to public health and the environment.

Regulatory Agencies in the United States

The biotechnology products and processes that are now being developed are diverse and their number is growing. Initially, recombinant DNA technology was limited to basic research laboratories where scientists attempted to increase the quantitites of existing compounds, usually available in trace amounts that are cumbersome and costly to isolate by conventional techniques. Their goal was to gain better insights into the molecular mechanisms governing living systems. By turning its attention to drugs, diagnostics and vaccine production, the pharmaceutical industry became the first industry to apply the products of biotechnology research. In addition to producing the same drugs more efficiently and at a fraction of the cost, they were able to focus their attention on new compounds and processes not available through conventional techniquesntly and at a fraction of the cost, they were able to focus their attention on new compounds and processes not available through conventional techniques. Most of the pharmaceutical research was performed in laboratories in compliance with the NIH Guidelines under the supervision of the Food and Drug Administration (FDA). Commercial development of biotechnology became more complicated in terms of the regulatory framework when products and processes were developed for use in the open environment. Such products range anywhere from growth hormones for domestic animals to microbial pesticides and improved genetically altered crops. The chemical industry is using biotechnology to develop new polymers, improved biodegradation products and genetically engineered microbes that can break down hazardous materials. Normally, these industries are regulated by several regulatory agencies, among them the U.S. Department of

Agriculture, the Consumer Product Safety Commission, the Occupational Health and Safety Administration in addition to EPA.

As a small number of products entered the market initially, the regulatory agencies, leading among them the FDA, did not consider it necessary to change their regulations. FDA issued a guiding document "Points to Consider" [2] and continued to regulate recombinant DNA products on a case-by-case basis. Other agencies have followed suit and adopted similar guidance according to their mandates. However, as the number of new products and processes continue to grow and the approval processes demands expedition, agencies must decide whether to alter their regulations, or continue to handle biotechnology products according to their old regulations. To ensure consistency, the Biotechnology Science Coordinating Committee (BSCC) was established in 1985 to investigate the possibility of common regulations or guidelines for the entire country and to handle regulatory differences among the various agencies. The more biotechnology expands and develops, the more it becomes obvious that this technology must be regularps, the more it becomes obvious that this technology must be regular cy in a consistent manner in order to assure the safety of public health and the environment without stifling technological innovation and competitiveness in the international arena.

The U.S. Environmental Protection Agency and Biotechnology

At present, biotechnology products that come under EPA regulations are divided into two categories. The first category includes chemicals and chemical processes that are regulated through the Toxic Substances Control Act (TSCA). Rather than declare genetically engineered microbes a separate category, EPA opted to declare DNA a chemical, and thus regulate recombinant DNA as a new chemical. The second category includes genetically engineered pesticides and microbes. These are regulated under the Federal Insecticide, Fungicide and Rodenticide Act (FIFRA). While under FIFRA a company needs to satisfy EPA's needs in order to receive a permit, under TSCA the burden of proof is on the Agency. If EPA does not respond within a certain period of time, usually three months, the company is automatically permitted to market its product. Once the product is on the market, EPA can still regulate it as a used chemical, although doing so requires a more complicated and lengthy process. Because it is more difficult a and more costly to regulate a used chemical than a new chemical, it is in EPA's interest to examine and approve the new compound as a new chemical within the provided time frame.

More than a dozen microbial pesticides are approved and registered with EPA. The organisms are marketed in several products for use in agriculture, forestry and insect control. These products were developed and tested in field trials on less than 10 acres, which were then exempted from FIFRA. Since then EPA has received several genetically engineered microbial pesticides. EPA now requires an experimental use permit for any field trial of genetically engineered microbe, its components, or products, regardless of the size of the plot. Only one genetically engineered product was submitted to EPA under TSCA. Conceivably, EPA could regulate biotechnology products and processes under other statutes such as the Clean Air Act, the Clean Water Act, the Resource, Conservation and Recovery Act (RCRA) and the Comprehensive Environmental Compensation and Liability Act (CERCLA). To date, none of these statutes have been applied, although as biotechnology gears towards scale-up and expansion, RCRA, the Hazardous Waste Act, and CERCLA, the Superfund Act, may be used to control waste products and spills. However, if any of the upcoming biotechnology products are shown to cause harm to human health and the environment, any of the above statutes can be activated at any time.

Risk Assessment of Genetically-Engineered Organisms

Given the variety of Federal statutes mentioned above, it is possible to regulate genetically engineered organisms under existing Federal statutes. In fact, despite several proposed bills in Congress, no new legislation for recombinant DNA has been introduced. Of greater concern to Federal agencies is how to assess the risks that this technology poses to human health and the environment.

As long as recombinant DNA research was limited to the laboratory, the NIH

Guidelines were adequate in providing four levels of physical containment and three levels of biological containment. These provisions assured the public that no organisms could escape the laboratory. The first risk assessment experiments were sponsored by the NIH in 1977 and were reported in a workshop entitled "DNA Experimentation with E. Coli K1". The proceedings of the workshop, which was held at Falmouth, Massachusetts, in June of 1977, were reported in the Journal of Infectious Diseases [3]. Most risk assessment experiments conducted for NIH at that time by the various laboratories were performed under much stricter conditions than are now required. As experiments with recombinant DNA molecules proved less harmful than previously thought, the Recombinant DNA Advisory Committee began to relax the NIH Guidelines.

The situation changed somewhat when biotechnology moved from the laboratory to the open environment. Agricultural applications of the technology began to emerge on the market as promising alternatives and further improvements to conventional technologies. Scientists began to examine frost and insect resistant plants, nitrogen fixing bacteria, and improved high yielding crops. In the chemical industry, researchers were looking at engineered organisms that can biodegrade hazardous spills and chemicals. Biotechnology required large fermentation facilities of which the products were not necessarily limited to the laboratory but were intended for use in the field and in urban environments.

In a collaborative agreement with the American Associated for the Advancement of Science (AAAS), EPA sponsored a series of workshops to assess environmental risks associated with recombinant DNA molecules in the environment. In the last of these workshops, which took place at Coolfont, West Virginia, in April of 1984, recommendations were made to pursue both environmental and health risk assessment research [4].

In the Spring of 1985, EPA acted on one of these recommendations and sponsored a conference entitled "Genetically Altered Viruses and the Environment". The conference, which took place at the Banbury Center in Cold Spring Harbor, New York, included virologists, molecular biologists, microbiologists, clinicians, public health officials and policy makers. Based on this conference a book the Banbury Report 22, was published in December, 1985 [5]. At the Banbury Conference some surprising information was presented on the stability of viruses over time and distance. For example, gastrointestinal viruses have been shown to survive in a water stream over a period of six months and a distance of 60 miles and were just as viable when put back into culture [6]. It was also reported that baculoviruses which are used as cloning vectors in genetically engineered microbial pesticides and fertilizers, have a shelf life of eight years [7]. Furthermore, novel recombinant viruses that one constructs can have unex pected biological properties under various environmental conditions and host range determinants [8]. Scientists agree, however, that most genetic manipulations yield less virulent organisms than those selected through a natural process. Yet given the unpredictable nature of biological molecules in the environment, it is difficult to predict what they may produce.

Three other conferences followed. The first one focused on "Risk Assessment of Genetically Altered Viruses" at the 1985 Annual Meeting of the AAAS [9]. The second one focused on "Genetically Predisposition to Disease" and was presented at the 1986 Annual Meeting of the AAAS [10]. A major part of the second Symposium focused on chemical and biological genotoxic agents which cause harm to humans and ecosystems. A third conference was held at the Banbury Center in the spring of 1986 on "Antibiotic Resistance Genes: Ecology, Transfer and Expression" [11] and was co-sponsored by EPA and National Science Foundation.

This series of conferences and symposia helped review important data that already existed among the various biomedical disciplines and interpret these from the point of view of environmental release. Based on information in these and previous meetings, a research agenda was drafted at EPA to define issues on the release of the recombinant DNA molecules into the environment. The more information was gathered on past examples of release, laboratory experiments and models, the more complex the research agenda became. Researchers realized that one set of experiments, or a single release, cannot adequately address the safety and risk concerns of biotechnology. The debates that surround release of genetically engineered organisms (12, 13, 14) made the agenda only more complicated. More data, further models and experiments were necessary in order to be able to conduct sound risk assessments.

Risk Assessment Research and the U.S. Environmental Protection Agency

The Coolfont Workshop in 1984 recommended some research directions. Since then EPA has sponsored several research projects testing the survival of genetically engineered organisms in soil and water. This research will be presented by others in this Workshop.

Let us focus on EPA-sponsored risk-assessment research. This research includes the risk assessment projects that I performed in collaboration with two academic institutions, the Department of Molecular Biology and Microbiology at Tufts University and the Department of Pediatrics and Physiology at Johns Hopkins University. At the time the research was performed, it was not possible to release genetically engineered organisms into the environment, not even for test purposes. Thus, two separate sets of experiments were design. The first set involved the testing of genetically engineered viral strains in the laboratory, while the other set of experiments tested natural bacteria and their plasmids in the environment. We are currently seeking permission to release the recombinant counterparts of the natural bacterial strains in the environment. Nevertheless, both the laboratory and the field experiments as presently designed have already yielded some interesting results.

The first project, performed at Johns Hopkins University, involved the risk assessment of genetically altered retroviruses in mammalian systems [15]. A cell culture system was designed to test the frequency at which retroviruses can alter infected cells in a deleterious manner. The work was designed to evaluate several aspects of retrovirus-cell interaction that present potential harm to humans and the environment.

We tested two systems to determine the frequency of retrovirus insertion mutagenesis for genetically engineered retrovirus RD-114 at a specific genetic locus. In the first system, we infected cells with retroviruses and then grew them in the presence of a chemical which kills any cell that has not lost the function of a target gene. Specifically, primary fibroblast cultures from normal males were used as the cells that were infected. The cells produce an enzyme hypoxanthine quanine phosphoribosyl transferase (HGPRT), the gene of which is on the X chromosome and thus is present in a single copy in males. Cells with HGPRT activity which are grown in the presence of 6-thioguanine (TG) incorporate this compound into DNA and are killed, but cells in which HGPRT activity is inhibited are resistant (TGr) and continue to grow.

The second system involves mutagenesis of fibroblast strains from individuals with hereditary bilateral retinoblastoma. Analysis of tumors from these individuals indicates that one homologue of chromosome 13 which contains the retinoblastoma (Rb) gene, is always absent from normal somatic tissues. Further, the chromosome which is present in tumor cells is always inherited from the parent who had retinoblastoma. Thus it appears that expression of the mutant Rb gene which leads to tumor formation is recessive, i.e. it is not expressed as long as the normal Rb gene is present in the cell. We hypothesized that inactivation of the normal Rb gene by retrovirus-mediated insertion mutagenesis might lead to transformation of cultured fibroblasts from retinoblastoma patients and result in the formation of foci on a monolayer of normal cells.

Our analysis of the two systems produced the following results. As can be seen in Table 1, in the HGPRT system, the frequency of spontaneous mutagenesis and of virus mediated mutagenesis is about the same. Moreover, the frequency of retrovirus-mediated insertion mutagenesis at the HGPRT locus is extremely low, on

TABLE 1

CELL LINE		NUMBER OF COLONIES	FREQUENCY (COLONIES/10^7 CELLS)
$343V^{+}2$		5	5.6
$343v^{+}2$		5	5.6
$343v^{+}$		2	2
$343v^{+}$		2	2
$344v^{+}$		6	4.8
$344v^{+}$		5	2.5
$345v^{+}$		1	1
	TOTAL =	26	3.2
$344_{CONTROL}$		6	7.5
$345_{CONTROL}$		5	6.25
	TOTAL =	11	6.9

FIGURE 1

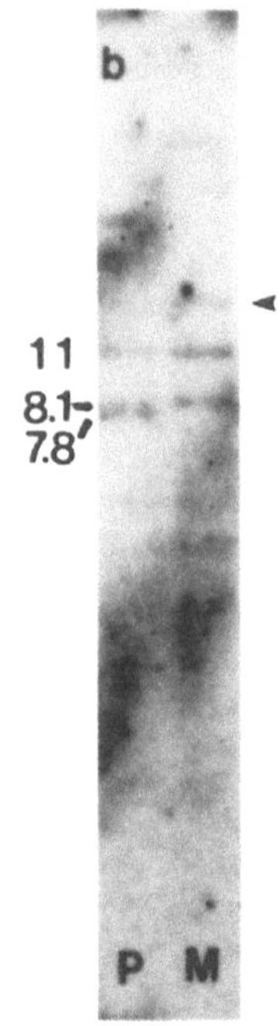

Figure 1: Analysis of retrovirus-mediated insertion mutagenesis. DNAs from the parental line 33V+ (P) and the TGr subclone 343V+1 (M) were digested with restriction endonuclease EcoRI and analyzed with a HGPRT-specific CDNA probe. Arrow indicates changes in HGPRT fragment size in the 343V+1 clone.

the order of one event per 1.6 x 10^8 integrations. Twenty-six TGr colonies arising from these experiments were clones and analyzed. Their DNA was isolated and they were examined for alterations in the structure of the HGPRT gene using Southern Blot analysis. Only one TGr clone, clone 3434^+, showed a structural change in the HGPRT gene which proved to be stable in two separate isolates of the same clone as shown in Figure 1 below.

Since a mutated gene at the HGPRT locus is an extremely rare event, we believe that the disruption in HGPRT gene seen in clone $343v^+$ is caused by infection with the retrovirus RD-114. Formal proof of inactivation by a retrovirus-mediated insertion mutagenesis would have required an expanded experiment yielding a viable statistical number of structurally-altered clones and molecular cloning of the mutated HGPRT gene, which was beyond the scope of our study. Had exposure to a recombinant retrovirus resulted in a large number of mutated colonies, there would have been sufficient concern to justify expanding the study to include analysis of the mutation at the molecular level. The second system with retinoblastoma cells was not capable of detecting retrovirus-mediated mutagenesis. Again, had results been positive, the experiemnt would have warranted a more extensive investigation. Although large scale experiments will be required to ensure the validity of the results, negative results thus far indicate that retrovirus-mediated mutagenesis is likely to result in, at most, the death of the affected cell and is of no consequence to the individ ual.

A second parameter measured the effect of retroviral oncogenes on human cells from normal individuals and from individuals with inherited predispositions to various forms of cancer. As shown in Figure 2, there was no appreciable difference in the frequency of focus formation obtained when cells from normal individuals or from individuals with any three different heritable cancer syndromes (intestinal polyposis, familial sarcoma, Wilm's tumor) when infected with Adeno-SV40 recombinant virus. However, the addition of other oncogenes, such as myc, ras, mos and sis, dramatically increased the frequency of focus formation by the Adeno-SV40

FIGURE 2

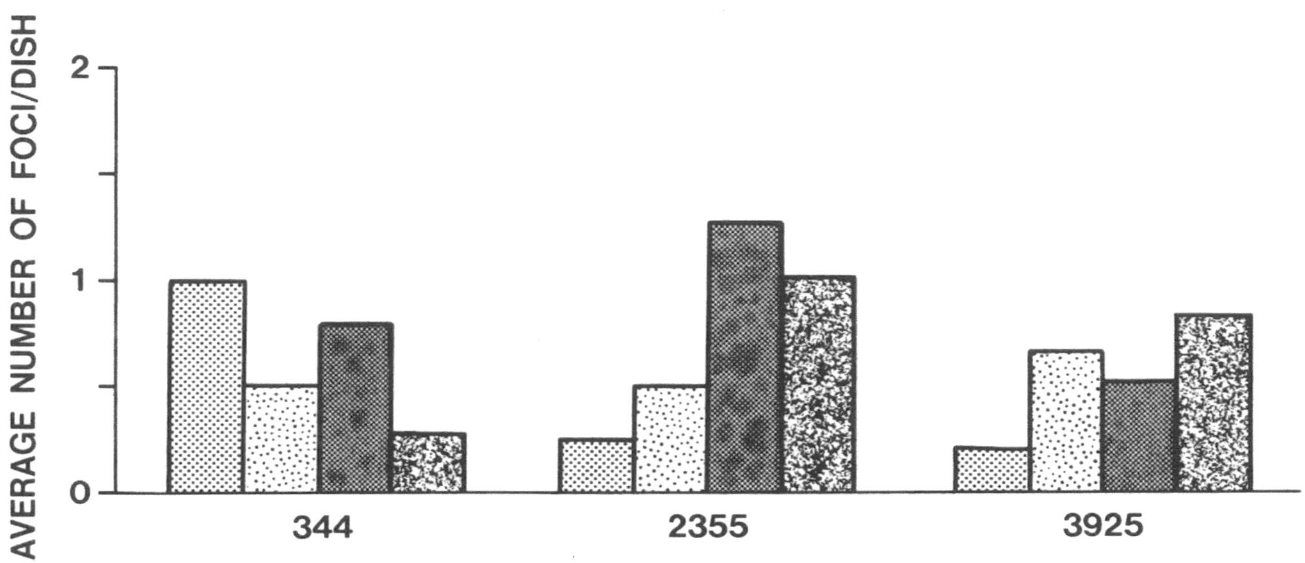

Figure 2: The genetic predispositions tested do not enhance the frequency of focus formation by Adeno-=SV40. Cells infected as described were scored for focus formation 2 weeks post-infection. Each point is the average number of foci from 4-5 dishes of infected cells plated 18 hours after infection. Each group of three represents, from left to right, initial infections using 0.5, 1, 2, or 4 x 10^6 PFU, respectively.

recombinant (Figure 3). This suggests that the "predisposing" cancers are caused by different genes which interact in different ways with external stimuli to produce transformation.

The negative results of both studies were encouraging in that they did not indicate direct mutagenic and/or transformation effects on cells caused by a recombinant virus. In order to validate these results they need to be repeated with other viral and bacterial strains and then tested in the open environment.

Since we would not release recombinant strains into the open environment, we designed experiments analogous to the laboratory research except we used natural bacteria, their derivatives and plasmids, for release into air (Marshall, B., Kamely, D. and Levy, S., manuscript in preparation, 15). Using aerosols as a means of dispersion, we have compared the survival of two E. coli strains, the laboratory strains X1666 and wild type SLH25, in the confined environment of a laboratory room and the semi-enclosed natural environment of a barn under ambient temperature and humidity conditions. The bacteria were dispersed in pairs with and without a Tn5-containing derivative of ColE1, the parental plasmid of commonly used cloning vehicles. Air and diverse types of surfaces were assayed for the organisms. In both environments, the number of bacteria declined rapidly within the first two hours. In the laboratory, the maximum survival time of any strain was 1-2 days; low numbers could be detected up to 20 days in the farm environment.

As indicated in Figure 4, airborne bacteria survived better in the barn environment than in the laboratory environment. While ambient temperatures of the two environments were similar, the relative humidity was much higher in the barn (61-94%) than in the laboratory (10-53%), which could account for the increased survivals. In the laboratory environment, the wild-type strain showed increase survival over X1666. However, die-off was too rapid to assess differences between plasmid-less and plasmid-bearing derivatives. In the barn, longer survival allowed evaluation of pairs of strains. These studies showed similar recoveries of the wild

FIGURE 3

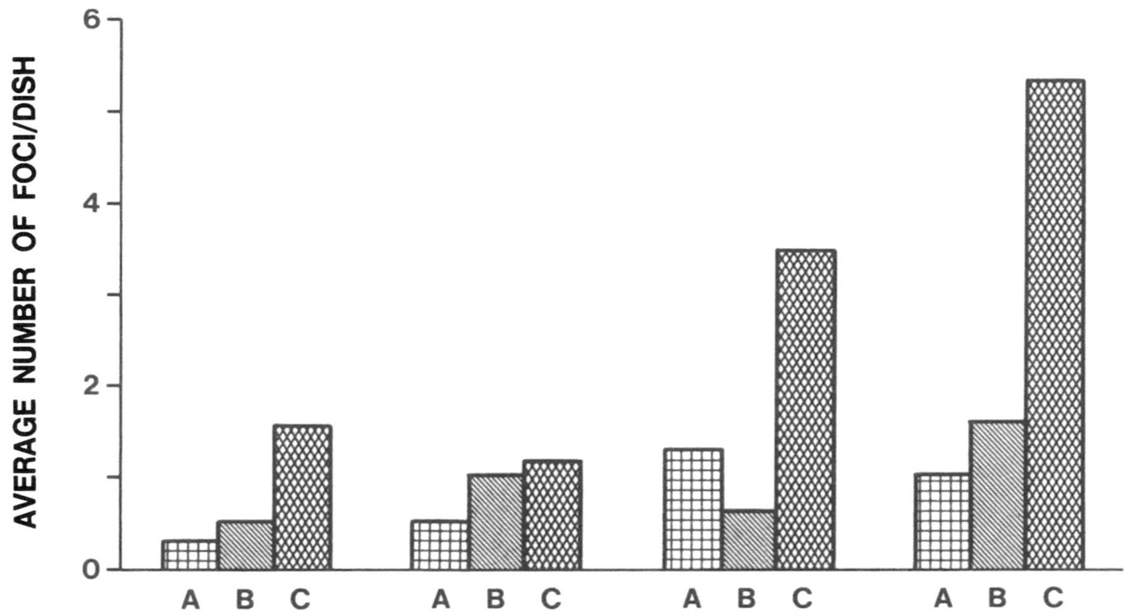

Figure 3: The presence of the ras oncogene alone is not sufficient to stimulate focus formation in Adeno-SV40 infected 2355 cells, but addition of ras, myc, mos and sis does enhance focus-forming transformation. a) 2355, fibroblasts from individual with a heritable adenomatosis; b) 2355 cells transfected with the ras oncogene; c) 2355 cells transfected with ras, myc, mos and sis oncogenes. The four groups represent four different multiplicities of infection.

FIGURE 4

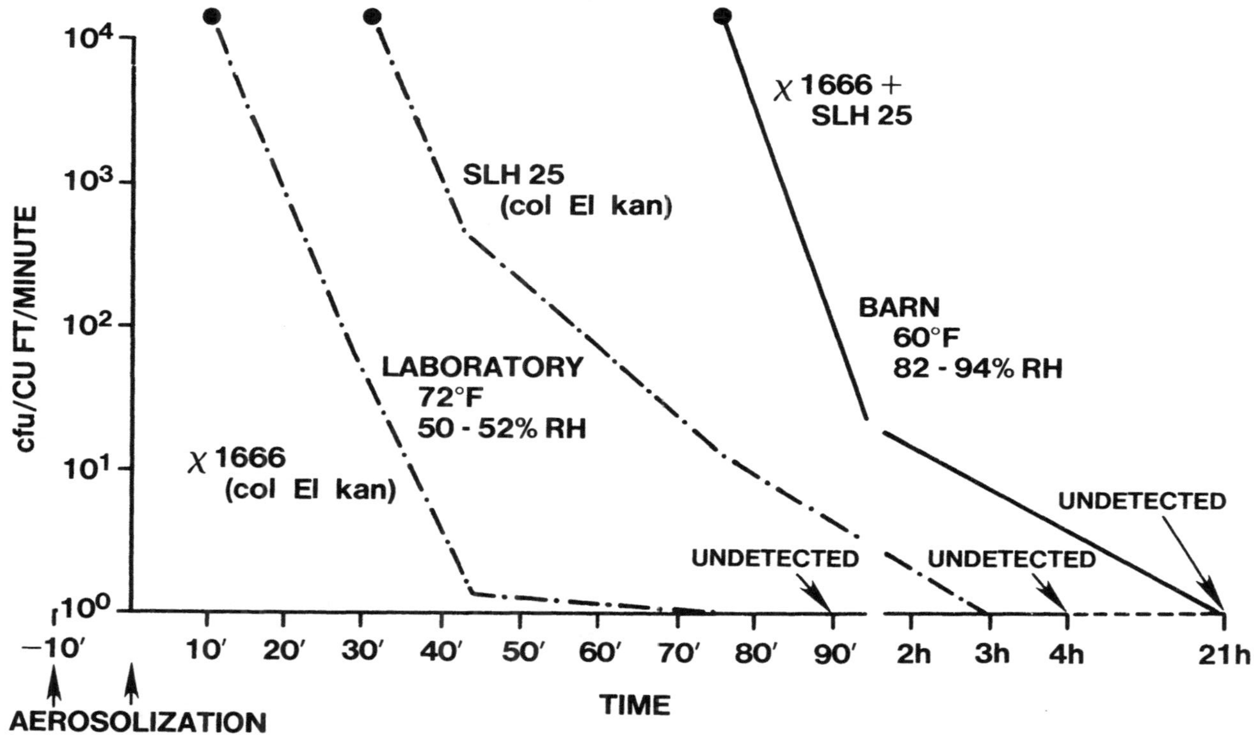

Figure 4: Survival of E. coli wild-type strain SLH 25 and laboratory strain X1666 in air. Airborne bacteria, recovered from air, survived better in the barn environment. Increased survival appeared dependent on the higher relative humidity.

type and the laboratory strain when retrieved from the air (see Figure 4) but higher recoveries for the wild type when measured on surfaces (not shown). Furthermore, the plasmid bearing derivative showed an apparent increase in survival over its plasmidless counterpart. In both environments the organisms survived on wood products such as paper and wood shavings.

These results argue that, if the laboratory strain X1666 exhibits weaker survival properties in both environments, a genetically-altered strain will be even weaker. As discussed earlier, opposing views on the subject have been presented and argued by scientists, environmentalists, and public representatives.

Conclusions

The above described experiments are only a sample of the risk assessment research that needs to be done in order to assure the safe release of genetically engineered organisms and their components into the environment. Although in our case the negative results are reassuring, they do not ensure the absence of deleterious effects that an altered gene may produce, whether in a virus, bacterium or a high organism. While debates over the safety of such releases have been going on ever since the first recombinant DNA strain was constructed, the public's concerns cannot be addressed until risk assessment experiments are actually performed both in the laboratory and in the environment. Many scientists and environmentalists fear, however, that once genetically engineered organisms are released into the environment, they can no longer be controlled.

How can we solve this dilemma? On the one hand, we have a technology with tremendous promise in the biomedical, agricultural and other industrial areas. On the other hand, we cannot advance because we fear biotechnology can have irreversible harmful effects which cannot be controlled beyond a contained environment.

There are two possible approaches. The first approach involves physical containment, i.e. a physical scale up of a gradual release from the laboratory to a greenhouse to a small field plot and finally to the open environment. Similar to the experiments at Tufts University which compared natural bacteria and their plasmids in the laboratory and in the barn, engineered bacteria can be introduced under controlled conditions and be carefully measured for several properties at each stage. Thus they can be stopped and retrieved from the restricted environment at any time. Only after the recombinant strains have been fully characterized and determined to be safe, they can then be released into the open environment.

The other approach entails the biological containment of an engineered organism. This has been successfully tried in gene therapy experiments in the laboratory using retroviruses as cloning vectors (17). Because retroviruses are also known to activate oncogenes, thus causing cancer in humans, scientists deleted the replicating portion of the viral vector. Thus the retrovirus can go through one cycle of replication, in which it can carry out its function, and then die off. Researchers are also working on biodegrading organisms which are engineered in such a way that they are dependent on a product they degrade. Once they complete their function, they can no longer survive.

These and other examples indicate that a step-wise controlled approach can be achieved to protect the public and at the same time advance biotechnology in a safe and controlled manner. Concerns raised by scientists, environmentalists, and the public requires the continued refinement of risk assessment models based upon statistically significant experimental evidence both in the laboratory and outside. Both science and the public will benefit from a cautious step-wise approach.

REFERENCES

1. National Institutes of Health Guidelines for Research Involving Recombinant DNA Molecules. Federal Register, July 7, 1976.

2. Statement of Policy for Regulating Biotechnology products. The Food and Drug Administration, Federal Register, Volume 51, No.123: 23309, June 26, 1986.

3. DNA Experimentation with E.coli-K1. Journal of Infectious Diseases, Volume 137, May, 1978.

4. Omenn, G. S. and Teich, A.H. Eds.: Biotechnology and the Environment: Research Needs. Noyes Data Corporation, Park Ridge, N.J. 1986.

5. Fields, B., Martin, M.A. and Kamely, D. Eds.: Genetically Altered Viruses and the Environment. Banbury Report 22, Cold Spring Harbor Laboratory, Cold Spring Harbor, New York, 1985.

6. Melnick, J. L. and Metcalfe, T.G.: Distribution of Viruses in the Water Environment. Banbury Report 22: 95-102,Cold Spring Harbor, New York, 1985.

7. Summers, M.D. and Smith, G.E.: Genetic Engineering of the Genome of the Autographa californica Nuclear Polyhedrosis Virus. Banbury Report 22: 319-331, Cold Spring Harbor, New York, 1985.

8. Hopkins, N.: The Genetic Basis of Leukemogenicity and Disease Specificity in Nondefective Mouse Retroviruses. Banbury Report 22: 195-210, Cold Spring Harbor, New York, 1985.

9. Risk Assessment of Genetically Altered Viruses. Abstracts of the Annual Meeting of the American Association for the Advancement of Science, 1985.

10. Genetic Predisposition to Disease. Abstracts of the Annual Meeting of the American Association for the Advancement of Science, 1986.

11. Levy, S.B. and Novick, R.P.: Antibiotic Resistance Genes: Ecology, Transfer, and Expression. Banbury Report 24, Cold Spring Harbor Laboratory, Cold Spring Harbor, New York, 1986.

12. Sharpless, F.E.: Regulation of Products from Biotechnology. Science, 235: 1329-1332, 1987.

13. Davis, B. D.: Bacterial Domestication: Enderlying Assumptions. Science, 235: 1329-1335, 1987.

14. Baltimore, D.: Setting the Record Straight on Biotechnology. Technology Review, 38-46. Massachusetts Institute of Technology Press. October, 1986.

15. Reeves, R. H., Littlefield, J. W. and Kamely, D.:Risk Assessment of Genetically Altered Viruses in Mammalian Systems. Manuscript in preparation.

16. Marshall, B., Kamely, D. and Levy, S.B.: Deliberate and Accidental Release into the Environment of Genetically Engineered Vectors and their Hosts. Manuscript in preparation

17. Anderson, W. F.: Prospects of Human Gene Therapy. Science, 226: 401-409, 1984.

PART IV

REGULATORY PERSPECTIVES

STATUS OF UNITED STATES GOVERNMENT REGULATORY ACTIVITIES FOR ENVIRONMENTAL APPLICATIONS OF GENETICALLY ENGINEERED ORGANISMS

John J. Cohrssen, Esq.
Attorney Advisor
Council on Environmental Quality
Executive Office of the President
Washington, DC 20503

INTRODUCTION

In this presentation, I will describe the US government's regulatory approach to the environmental applications of genetically engineered organisms and the assumptions underlying the US approach. The assumptions are important since they determine what research and products are regulated. Assumptions also determine the outcome of the risk assessments.

The US government has taken a very active role in relation to biotechnology. US Dollars have paid for a major portion of the increased knowledge that has enabled the development of rDNA and other new technologies based upon molecular biology. The US has also established probably the most complete, and maybe most complex, regulatory framework.

The US has sought to assure public confidence by establishing regulations based on scientific knowledge, as well as speculative concerns. We have had substantial experience with genetically manipulated organisms including rDNA organisms that provides the scientific basis for risk assessment. To the extent we find that regulations are based upon concerns that are not scientifically plausible, unrealistic demands will be placed upon risk assessors and risk managers. Fortunately, the US regulatory framework is sufficiently flexible to allow refinement and narrowing of regulatory categories.

Once before, the US established rules because of concerns related to genetic engineering. This was some fifteen years ago when hypothetical concerns about recombinant DNA manipulation of cancer causing genetic material led to the development of the NIH Guidelines for Research with Recombinant DNA Molecules. By using methods of risk assessment, the NIH Recombinant DNA Advisory Committee (RAC) considered the hypothetical hazards

NATO ASI Series, Vol. G18
Safety Assurance for Environmental Introductions
of Genetically-Engineered Organisms
Edited by J. Fiksel and V.T. Covello

associated with various experiments, and over the years developed internationally accepted standards for various levels of review by research institutions and agencies funding research. This was "self-regulation" by scientists and research funding agencies. NIH review of individual experiments on a case-by-case basis was replaced by "compliance standards" developed by recognized scientists on the NIH RAC and RAC Working Committees. The eventual reduction in requirements by the NIH mirrored public confidence in the safety of rDNA technology. I believe that we will see a similar experience in environmental applications of genetically engineered organisms.

RISK ASSESSMENT ASSUMPTIONS

The US Coordinated Framework for the Regulation of Biotechnology published on June 26, 1986 relies on various assumptions with respect to both defining the categories of regulated products and the levels of regulatory review. As the scientific issues related to the assumptions become better understood, categories of regulated products and approaches to regulatory review may be modified.

We need to be sure that assumptions are scientifically defensible in their meaning, approach, and use. As I said earlier assumptions determine the outcomes of risk assessments. Assumptions can also affect the perception of risks

Assumptions are also implicit in the language we use. Consider the familiar bad example of the term "deliberate release." The term circumscribes the areas of concern for this workshop. The term deliberate release may illustrate the type of conceptual problems that we can face and demonstrates the need for meticulous description.

We must remember that biotechnology is actually a collection of very different biological processes or enabling techniques forced under one very large umbrella. In a similar way, release may equate a large number of very different types of biological introductions in various environments.

Let me raise two issues: First, the pejorative nature of the term "deliberate release" or even "release" that may affect risk perception; second, what type of definition of release do we need.

In my view the original choice of the term deliberate release was unfortunate and wrong because the phrase conveys the image of deliberatively freeing a dreaded organism that will go off, out of control, and cause harm. This implies possibly doing something damaging to the environment. The word deliberate implies a malicious intent. Moreover, "release" means to unfetter, free or let go. That is a problematic act. We don't speak of deliberately releasing seeds in the ground, or fertilizers on top of the ground or pesticides in the air. We plant seeds, we apply fertilizer and pesticides. We don't deliberate release farm animals. We don't release vaccines. The point is that the routine introduction of each of these products is conceptually different from the concept of deliberate release expressed in the NIH Guidelines. (As an aside, I would like to mention that in our legislature, bills are introduced, not deliberately released. Some might think that our legislation is in fact deliberately released.) The drafters of the NIH Guidelines were, of course, concerned with dangerous pathogens such as cancer causing pathogens, a concern that years later was considered to be very exaggerated. My point is simply that we must take care to be very precise in our concepts.

The June 26, 1986 Coordinated Framework requested comments on the definition of release because the term could determine regulatory categories of what should and what should not receive a priori review. We received relatively few comments. In its section of the Coordinated Framework, the EPA, as an interim approach, used the concept of "outside of physical containment." The Animal and Plant Health Inspection Service also defines release as removal out of a contained laboratory or contained greenhouse. Thus for APHIS, permit requirements are triggered when an organism achieved through rDNA techniques and containing genetic material from a plant pest is released. EPA recognized the limitations of its definition and has been seeking a better definition through the use of experts. The NIH RAC also has been scrutinizing definitions and recently agreed to remove self cloning experiments involving environmental applications from its deliberate release requirements.

Since we are seeking to describe the real hazards associated with environmental applications, we need to be certain that a unitary concept of release does not lead to overly inclusive, and therefore, inappropriate categories. We need to satisfy the scientists who would object to an unsuitable lumping together of the very different types of concerns and hypothetical hazards for many diverse organisms and situations. We also need to satisfy the regulatory requirements for administrative simplicity.

I would like to now turn to the following concepts written in the June 26 policy, the US Regulatory Framework and define categories of regulated products and levels of regulatory review.

We make true statements:

o Organisms may reproduce, spread, and become established.

o Organisms may be pathogens causing disease in humans, plants, animals, and other microbes.

o Genetically enhanced microorganisms, by any technique, have potential to exhibit new traits or characteristics.

o New traits in an organism may imply hazards.

We do not include certain true statements such as the fact that organisms die, and new varieties have difficulty competing and surviving.

We make assumptions that have limited scientific support. We make assumptions for pathogens:

o A non-pathogenic organism that is manipulated to contain some amount of genetic material from a pathogen or plant pest may exhibit pathogenic or plant pest characteristics, except when the new genetic material consists of certain well-characterized, non-coding regulatory sequences.

o There is increased uncertainty about behavioral changes associated with genetically engineered organisms containing genetic material from pathogens; all should be reviewed by a national agency prior to release.

o All pathogens tend to be equated in terms of risk to health or the environment.

We make <u>assumptions for intergeneric combinations of microorganisms</u>:

o Inter-generic combinations (combinations from source organisms of different genera), but not intra-generic combinations (source organisms from the same genus) are sufficiently likely to result in new combinations of traits that should be given special attention. All new traits tend to be equated with a risk.

- Combinations of genetic material from microorganisms from different genera are more likely to result in new combinations of new traits than combinations of genes from microorganisms within the same genus.
- While genetic exchange occurs naturally and somewhat commonly among many microorganisms, it is more likely to occur in nature within a single genus than across many different genera.
- Genus designations provide a practical criterion for administrative and regulatory purposes.

o Certain inter-generic combinations do not require review when the genetic material added to the recipient microorganism consists only of well-characterized, non-coding regulatory regions.

a. The resulting organisms do not possess new combinations of traits.

b. They exhibit quantitative changes in preexisting traits

We make <u>assumptions for nonengineered pathogens</u>:

o Nonengineered indigenous pathogens are of lesser risk and do not require review by national authorities until they are used in large scale applications (greater than 10 acres).

- Ample experience indicates that nonengineered, indigenous pathogens are sufficiently well controlled by natural mechanisms in small-scale environmental applications.

o Nonindigenous microorganisms are more likely than indigenous microorganisms to exhibit traits new to an environment into which they are applied.

o Indigenous microorganisms are more likely to be subject to natural limitations.

An assumption that is sometimes used, and often vigorously challenged, is the similarity of the introduction of enhanced organisms to the introduction of nonindigenous or exotic organisms. There is a general recognition that a small number of introductions of nonindigenous organisms have sometimes led to untoward environmental consequences. However, there is no experiential evidence to support the assumption that a genetically engineered organism will be analogous to exotic species. The reason is that engineered organisms are almost congruent to non-engineered. Their introduction is analogous to the introduction of a new plant variety. Accordingly, only if the engineered host organism is nonindigenous, would there appear to be reason for concern.

Also, challenged, are concerns about the effects of horizontal gene transfer that are believed to have been greatly exaggerated. Although such exchanges are known to take place, there is insufficient scientific evidence to assume that the transfer of recombinant material will take place and have an untoward effect. Our knowledge regarding these assumptions is rapidly increasing. Accordingly, regulatory categories will be narrowed, and risk assessments will be more precisely done. Questions continue to arise regarding the required level of review by a regulatory agency for particular classes of experiments or products. There has been no evidence of hazards attributable to rDNA manipulations after some fifteen years of experience with rDNA molecules. And, experience with microorganisms modified by traditional techniques has not raised the concerns expressed regarding genetically engineered organisms even though these traditionally modified organisms are considered to be much less precisely engineered.

Where is the appropriate place in our assumptions for consideration of the margin of safety that is imparted to the environmental introductions of microorganisms because of the enormous buffering capacity in the microbial world? Introductions occur all the time by natural events through mutation and genetic exchange, horizontal gene transfer, and the like, as well as by natural perturbations. Our ability to easily make undesirable perturbations and pivotal changes may be greatly exaggerated. How can we best weigh the fact that the environment is complex and dynamic, but fundamentally stable, and all but a very small number of introductions tend have no significant effect. Possible hazards of an introduction need to be considered against this vast background activity, its overall level of stability, and our ability to control experimental conditions and thereby minimize the opportunity for untoward events.

The challenge is then to continue to refine the assumptions that require regulatory risk assessments, and the extent or depth of assessment that is necessary.

THE COORDINATED FRAMEWORK

The June 26, 1986, final Coordinated Framework for the Regulation of Biotechnology established categories of genetically engineered organisms that require federal review by one or more US agencies. We believed that the "new biotechnology" should be regulated the same as "old biotechnology"; and our existing laws were appropriate and sufficient to manage research and products. Of course, some regulatory "tinkering" which we call "rulemaking" was needed. Even though there may be significant differences in the risks that attach to various types of environmental applications, we believed that a national program should attempt to have uniform definitions and approaches. To avoid confusion, and to simply regulation, we sought to identify a single or lead agency for all applications, although there are unavoidable variations because of each agency's legal requirements.

I should mention that because the term biotechnology is ubiquitous, the laws that regulate old biotechnology span different approaches and indeed different concepts of risk, or even concepts of acceptable risk. In practical terms, this means that our "Coordinated Framework" could easily confound and confuse persons not generally familiar with our complex US regulatory regimes.

I will summarize the basic Coordinated Framework requirements for environmental applications. I will not describe the framework regulations that relate to manufacturing processes and their products, but will make some reference to other non-environmental products such as vaccines and foods.

Product Regulation

There is a fundamental principle that I must emphasize regarding products. Agencies involved with regulating agricultural and environmental products have had extensive experience with the safe use of products that involve applications in the environment. By the time a genetically engineered product is ready for commercialization, it will have undergone substantial review and testing during the research phase, and thus, information regarding its safety, and anticipated benefits should be available. The manufacture will be reviewed by EPA and USDA in essentially the same manner for safety and efficacy as products obtained by other techniques. Typically, the agencies rely upon qualitative risk assessments. The regulatory scheme for products is described in Chart I Coordinated Framework -- Marketing Approval of New Biotechnology Products Applied in the Environment.

CHART I — COORDINATED FRAMEWORK — APPROVAL OF NEW BIOTECHNOLOGY PRODUCTS APPLIED IN THE ENVIRONMENT

Subject	Responsible Agency(ies)
Foods/Food Additives	FDA*, FSIS[1]
Human Biologics	FDA
Animal Biologics	APHIS
Plants and Animals	APHIS*, FSIS[1], FDA[2]
Microbial Pesticides	
All	EPA*, APHIS[3]
Other Microbial Products	
Intergeneric Combination	EPA*, APHIS[3]
Intrageneric Combination	
Pathogenic Source Organism	
1. Agricultural use	APHIS
2. Non-Agricultural use	EPA*[4], APHIS[3]
No Pathogenic Source Organisms	EPA Report
Nonengineered Pathogens	
1. Agricultural Use	APHIS
2. Non-agricultural Use	EPA*[4], APHIS[3]
Nonengineered Nonpathogens	EPA Report

* LEAD AGENCY

1 FSIS, Food Safety and Inspection Service, under the Assistant Secretary of Agriculture for Marketing and Inspection Services is responsible for food use.

2 FDA is involved when in relation to a food use.

3 APHIS, Animal and Plant Health Inspection Service, is involved when the microorganism is plant pest, animal pathogen or regulated article requiring a permit.

4 EPA requirements will only apply to environmental release under a "significant new use rule" that EPA intends to propose.

Jurisdiction over the various biotechnology products is determined by their use, as has been the case for traditional products.

The Animal and Plant Health Inspection Service, (APHIS) reviews animal biologics, animal pathogens and plant pests including "regulated articles", i.e., certain genetically engineered organisms containing genetic material from a plant pest. An APHIS permit is required prior to the importation, shipment or release into the environment of regulated articles, or the shipment of a plant pest or animal pathogen. APHIS will place the burden on applicants to determine--with the aid of a list of plant pests--whether the genetic material inserted into the host is a plant pest. Likewise, EPA will provide a definition to determine when an intrageneric microbe is a "pathogen."

Microbial pesticides are reviewed by EPA, with APHIS involvement in cases where the pesticide is also a plant pest, animal pathogen, or regulated article. (FDA may become involved in implementing pesticide tolerances for foods.)

For other products, jurisdiction depends on the characteristics of the organism as well as its use. Intergeneric microorganisms, transgenic combinations, microorganisms need EPA approval under Premanufacturing Notice (PMN) requirements, with APHIS involvement in cases where the microorganism is a also a regulated article.

"Intrageneric combinations" are those microorganisms formed by genetic engineering other than intergeneric combinations. APHIS has exclusive jurisdiction for these when there is a pathogenic component of an organism, and the microorganism is used for agricultural purposes. If a microorganism is used for nonagricultural purposes, then EPA has jurisdiction, with APHIS involvement in cases where the microorganism is also a regulated article requiring a permit. Although intrageneric combinations with no pathogenic source organisms are under EPA jurisdiction, EPA will only require an informational report.

Importation and shipment of all plant pests and animal pathogens fall under APHIS jurisdiction. Those items that are for a nonagricultural use may also come under EPA jurisdiction. Nonengineered nonpathogenic microorganisms are under EPA jurisdiction which will require an informational report.

Research

The Coordinated Framework for the regulation of biotechnology expands requirements for the conduct of research first developed in the NIH Guidelines describing the manner in which research with organisms derived by rDNA techniques should be conducted. The guidelines prescribe the conditions under which institutions which receive NIH (and other governmental) funds must conduct experiments. For a very small category of NIH funded experiments including environmental release, the guidelines require that the NIH Director approve each experiment on an individual basis. For each of these experiments, the RAC conducts a scientific review with an opportunity for public comment, and makes a recommendation to the NIH Director. As research experiments using recombinant organisms have expanded out of the biomedical area to environmental applications both agricultural and nonagricultural, other agencies have become involved, with shifting of responsibility for research approval to NSF, USDA and EPA. These other agencies' policies build, in part, on the NIH Guidelines and NIH experience.

It should be noted that not all experiments involving the environmental application of genetically enhanced organisms require prior federal approval. For certain categories of microorganisms including those modified by traditional genetic modification techniques, there is a substantial body of research indicating low risk for environmental experiments.

Chart II -- Coordinated Framework -- Approval of New Biotechnology Research Applied in the Environment shows which agency has responsibility for a particular experiment. If more than one agency has potential jurisdiction, one agency has been designated as the lead agency and it is marked with an asterisk on Chart II. The lead agency designation depends on which research agency is funding the research (e.g. NIH, S&E, or NSF) or which regulatory agency reviews specific purpose research (e.g. pesticides). In the chart and in this discussion, the authority refers to approval of the actual execution of experiments and not to their funding.

CHART II--COORDINATED FRAMEWORK-- APPROVAL OF NEW BIOTECHNOLOGY RESEARCH APPLIED IN THE ENVIRONMENT

Subject	Responsible Agency(ies)
Foods/Food Additives, Human Drugs, Medical Devices, Biologics, and Animal Drugs	
1. Federally Funded	FDA*, NIH guidelines & review
2. Non-Federally Funded	FDA*, NIH voluntary review
Plants, Animals and Animal Biologics	
1. Federally Funded	Funding agency*1, APHIS2
2. Non-Federally Funded	APHIS*, S&E voluntary review
Microbial Pesticides	
Genetically Engineered	
Intergeneric	EPA*, APHIS2, S&E voluntary review
Pathogenic Intrageneric	EPA*, APHIS2, S&E voluntary review
Intrageneric Nonpathogen	EPA*, S&E voluntary review
Nonengineered	
Nonindigenous Pathogens	EPA*, APHIS
Indigenous Pathogens	EPA*3, APHIS
Nonindigenous Nonpathogen	EPA*,
Other Microbes Released in the Environment	
Genetically Engineered	
Intergeneric Organisms	
1. Federally Funded	Funding agency*1, APHIS2, EPA4
2. Commercially Funded	EPA, APHIS, S&E voluntary review,
Intrageneric Organisms	
Pathogenic Source Organism	
1. Federally Funded	Funding agency*1, APHIS2, EPA4
2. Commercially Funded	APHIS*2, EPA (*if non-agricul.use)
Intrageneric Combination	
No Pathogenic Source Organisms	EPA Report
Nonengineered	EPA Report*, APHIS2

* LEAD AGENCY

1 Review and approval of research protocols conducted by NIH, S&E, or NSF.

2 APHIS issues permits for the importation and domestic shipment of certain plants and animals, plant pests and animal pathogens, and for the shipment or release in the environment of regulated articles.

3 EPA jurisdiction for research on a plot greater than 10 acres.

4 EPA reviews federally funded environmental research only when it is for commercial purposes

The NIH RAC has expressed its intention to incorporate into the NIH Guidelines, appendices applying to planned applications of microorganisms, plants and animals in order to provide guidance for research on rDNA containing organisms. These guidelines specify what type of review procedures are required for specific categories of experiments. Some individual experiments require prior approval by the agency that provides institutional support. For those experiments that require agency approval, advisory committees at NIH, S&E, and NSF, composed primarily of nongovernment scientists, may be asked to provide expert review. In addition, research on agricultural products will come under APHIS permit requirements if a regulated article, plant pest, animal pathogen is involved. An APHIS permit may be required prior to the shipment (movement) or release of a regulated article, or the importation or shipment of a plant pest or regulated article used in any research experiment.

EPA will review all proposed environmental applications of all microbial pesticides regardless of whether research is federally funded or not. EPA will regulate research under a two level review system (based upon assumptions about the risks posed by various types of microorganisms) with lesser data requirements imposed for level I reporting and more comprehensive review for level II. For the "other uses" category from Chart II (research involving nonpesticidal microorganisms applied into the environment), jurisdiction for release may be under S&E, NSF, APHIS, or EPA depending on: (i) the source of funding, and (ii) the purpose of the research and (iii) the characteristics of the genetically engineered microorganism. Thus, federally funded research conducted for an agricultural use will require adherence to NIH guidelines and approval of certain experiments by S&E or NIH depending on which is the funding agency. EPA will review commercial research. APHIS's jurisdiction applies to issuing permits for regulated articles, plant pests, or animal pathogens. EPA will require an informational report for nonengineered microorganisms released into the environment, with APHIS involvement for the review of plant pests or animal pathogens.

There may be situations where one agency may choose to defer to, or ask advice from, another agency. If experiments requiring NIH, NSF or S&E review/approval are submitted for review to another agency, then NIH, NSF, or S&E may determine that such review serves the same purpose, and based upon that determination, notify the submitter that no NIH, NSF, or S&E review will take place, and the experiment may proceed upon approval from the other agency.

NON-REGULATORY ASSESSMENTS

I would like to mention briefly non-regulatory uses of risk assessment. The fact is that researchers and manufacturers, as well as the government, routinely conduct various assessments related both to the laboratory use and to environmental use of microbes. Safety assessments and risk assessments are routinely relied upon by industry and governments for the setting of priorities. Some type of assessment is an essential part of all activities related to research and development, especially when pathogens or toxic chemicals are used.

One reason for interest in safety assessment and risk assessment is the strong anxiety in the US regarding tort liability. The US system of law might allow a recovery in the situation where the manufacturer or researcher failed to consider the risks involved in an environmental application if some type of injury resulted; both compensatory and punitive damages could be assessed. The legal systems in other countries may not provide similar relief. A possible second reason is the constellation of federal and state criminal laws that could be applied in the event that one failed to reasonably assess microorganism characteristics and there was injury to health or the environment. For example, there are federal and state plant pest acts that can provide criminal sanctions for the unapproved movement or introduction of plant pests. Moreover, other criminal laws could lead to prosecution and possible fine or imprisonment for those that negligently allowed injury to persons or the environment.

Most situations do not require a risk assessment because safety standards or established practices dictate the way in which research or manufacture need to take place. Thus, for example, the NIH Guidelines establish the manner--the levels of physical and biological containment--in which certain laboratory experiments must take place. Were an untoward result to occur after a researcher failed to comply with the guidelines, the non-compliance could be used as a basis for demonstrating tort or criminal liability. The NIH Guidelines although written for federal grantees have also set a nationwide standard for proper conduct of research with rDNA molecules. Insurance companies require that research be conducted according to these standards because of the recognition that the courts would consider the NIH Guidelines a legal standard for the proper conduct of research.

The establishment of a standard implies that a determination has been made of an acceptable upper bound of probable risk, and that compliance with the standard will reduce the level of risk to an acceptable level. In regulatory situations, a standard may also constitute a trigger that then requires further regulatory action, usually *a prior* approval of a particular activity.

CONCLUSION

To summarize, we have made significant progress in developing a US framework for defining and regulating the environmental application of those products of biotechnology that are of concern. The framework seeks to protect health and the environment, but the framework is flexible enough to allow rapid modification to keep pace with rapidly evolving scientific developments. By design the framework is somewhat conservative, and somewhat over regulatory, requiring more rather than less regulatory requirements, but we expect that deregulatory refinements will follow. Some refinements may, in fact, emerge from deliberations at this NATO workshop as we examine the assumptions underlying risk assessment.

I would like to make a final point regarding the responsibility of the risk assessor. Much has been written about the fact that regulation and fear of liability litigation have tended to have a serious negative impact upon innovation, and scientific exploration. While this is not the topic of this workshop, it is important to remember that the tools that we are developing will be broadly used and thus, should be meticulously defined, scientifically sound, and judgment neutral. Our job is to responsibly move from "worst case" conjecture to approaches supported by credible science.

The Oversight of Planned Release in the UK

Brian P. Ager
Secretary, Advisory Committee on Genetic
Manipulation, Health and Safety Executive, Baynards House
1 Chepstow Place, London W2, 4TF, UK

Summary

This presentation outlines the arrangements for the oversight of planned release in the UK. These have been influenced by the recently issued major international study set up by the Organization for Economic Cooperation and Development (OECD) on recombinant DNA safety considerations. The Advisory Committee on Genetic Manipulation (ACGM) has been active in drawing up guidelines in a number of areas including planned release into the environment. ACGM has provided the focus for reviewing the first release in the UK of a genetically manipulated micro-organism.

UK Regulations for Generic Manipulation

The current UK health and safety requirements applicable to genetic manipulation are based on general legislation, specific notification regulations, published guidelines and on-site inspection of health and safety standards by the Health and Safety Executive (HSE). The requirements of the Health and Safety at Work etc Act 1974 cover genetic manipulation techniques and their application. The Act places a general duty on the employer requiring,

> "the provision and maintenance of a working environment for his employees that is, so far as is reasonably practicable, safe without risks to health and adequate as regards facilities and arrangements for their welfare at work."

The employer is also charged with a duty to avoid exposure of those not in his employment, including the general public, to risks.

It should be noted that these general duties are qualified by the phrase:

> "..... so far as is reasonably practicable ....."

In essence, the employer must make a cost-risk analysis and assess, on the one hand, the risk of the work and, on the other hand, the difficulty and expense involved in avoiding that risk. Thus greater risks require greater precautions. The theme of the Act is one of getting the right balance with a clear emphasis on self-regulation.

Aside from the general requirements of the Health and Safety at Work etc Act, the Health and Safety (Genetic Manipulation) Regulations 1978 (currently under review) require the notification to HSE of intention to carry out genetic manipulation as defined and the provision of details of individual experiments. Individual proposals submitted under these Regulations are considered by HSE and circulated to members of ACGM. Generally speaking, ACGM and its HSE Secretariat concentrates on the biological aspects of a particular proposal and HSE's Specialist Inspectors deal with physical containment aspects during on-site inspection.

There are various guidelines published by HSE indicating how an employer may discharge

NATO ASI Series, Vol. G18
Safety Assurance for Environmental Introductions
of Genetically-Engineered Organisms
Edited by J. Fiksel and V.T. Covello

his responsibilities under the Health and Safety at Work etc Act. For genetic manipulation detailed guidelines drawn up by the ACGM are available from HSE. In effect, these indicate how to achieve the right balance between the level of risk and the cost of taking precautions to avoid that risk under the general requirements of the Health and Safety at Work etc Act. It should be noted that such guidelines are not themselves inflexibly enshrined in regulations.

OECD Study on Safety and Regulations in Biotechnology

The OECD report "Recombinant DNA Safety Considerations"[1] was published towards the end of 1986 and I would like to highlight two points about this study.

1. The Group strongly believed that the establishment of an internationally agreed framework is an important step towards facilitating the development of biotechnolgoy, with its increasing promise of considerable benefit, whilst helping to ensure that due regard is paid to any potential concerns.

2. Although recommendations made by the OECD are not binding on Member Countries, there is little doubt that the report of this major international study will have considerable influence on the development of regulations and guidelines worldwide. Already we have seen the influence of the OECD report in several countries 2,3,4,5.

There is also a third point to make. This concerns public perception of risk and achieving the right balance between guidelines/regulations that are too harsh, and that therefore stifle progress, and those that are ineffective or too relaxed and do not therefore instil public confidence. There can be little doubt that without a good measure of public confidence in the way that we oversee risks -- real or potential -- new technologies such as biotechnology will not develop as they might. Without a framework for risk assessment (and without the right balance to that framework) we may not see biotechnology fulfill its many promises. The OECD study gives us a vital lead in our attempts to get the balance right.

UK Advisory Committee on Genetic Manipulation

In the UK oversight of genetic manipulation centres on ACGM and the HSE. ACGM was set up in 1984 to replace the Genetic Manipulation Advisory Group (GMAG). ACGM advises the Health and Safety Commission and the HSE, which provides the Secretariat, and government departments where necessary. It consists of an independent Chairman, representatives of employers and employees together with scientific and medical specialists.

Planned Release

At its first meeting ACGM recognized that the issues raised by planned release should be considered as a priority and guidelines[6] were issued in April 1986. The guidelines were drawn up by a specialist working group of ACGM together with representatives from HSE and government departments taking account of international activities, notably the OECD study.

In essence ACGM's guidelines require prior notification of individual proposals following local risk assessment. The guidelines identify a number of factors that should be used in such initial risk assessment. Clearly it is important that the scientific basis on which release proposals are judged is open to review as experience and the results of research accumulate and ACGM has established a Planned Release Sub-Committee to advise on individual proposals and to review the guidelines as necessary.

The guidelines do not establish an approval or licensing system and the notification requirement is not at present a statutory obligation. However, there are proposals for the review of the current Regulations and it is intended that notification of release projects be made statutory. What the guidelines do is to put in place a framework for the assessment of planned release on a case by case basis by a national expert committee, HSE and other relevant government agencies (Ministry of Agriculture, Fisheries and Food, Department of the Environment, Department of Health and Social Security and the Nature Conservancy Council).

There are also notification requirements under the Food and Environmental Protection Act 1986 for pesticides and under the Plant Health Act 1967 that may be relevant for some release applications. This topic is being studied by the UK's Royal Commission on Environmental Pollution and its report is expected in late 1987.

This framework established by ACGM has been exercised several times already. This conference will be aware for example that in late 1986 a genetically manipulated micro-organism was released in the UK. The project involved the release by the Natural Environment Research Council of a genetically marked baculovirus Autographa californica nuclear polyhedrosis virus (AcNPV) in a small "cabbage patch" ecosystem[7]. The virus was marked with a non-coding synthetic oligonucleotide sequence in an intergenic region of the viral genome.

Such viruses in their non-manipulated form have been used for some time as insecticides. This project is the first step in a programme for making "custom designed" viral insecticides with limited host range and limited persistence. The object of this first study was to measure the ability of the virus to persist on foliage and in soil in the ecosystem under study.

This initial field trial was given detailed review by ACGM's Planned Release Sub-Committee. The conclusion reached was that no objection could be seen to the genetic manipulation aspects of the project given the non-functional nature of the insert, the very limited scale of the project and the fact that the release was to take place under partially contained conditions. The virus was released in preinfected caterpillars of Spodoptera exigua in a netted and fenced compound to prevent dispersal of the host and interaction with other insects and predators.

The genetic manipulation provoked relatively little concern during the review. However the use of AcNPV itself gave rise to some debate and extensive heat range testing in Lepidoptera and Hymenoptera was carried out in agreement with the Nature Conservancy Council.

The ACGM guidelines and the Sub-Committee provided the focus for the review process and was followed by clearance under the UK Pesticide Safety Precaution Scheme (now superseded by the Control of Pesticides Regulations 1986).

Future Developments

It was recognised by ACGM that the release guidelines would need to be reviewed in the light of experience and as the results of research accumulate. Also against this background the process of review should begin to move away from a concensus of experts and other relevant groups towards an established risk assessment scheme. This is analogous to the development of the UK risk assessment scheme for genetic manipulation in the laboratory[8] which was a vital element in the move away from the earlier more general risk assessment principles and from a national case-by-case review of every experiment. So far in the UK we have seen a handful of release proposals involving limited and very small scale trials. However it is reasonable to expect that over the coming years the pace will quicken and the scale of releases will increase. There is therefore a need to make the review process more efficient (but no less effective) in a way that may eventually allow, with confidence, movement away from a case-by-case review for every proposal. The initial case-by-case consideration by national watchdog bodies such as ACGM should also lead towards greater efficiency of review as "case law" is established.

So far this type of review has been based on the principles and "points to consider" laid down in the OECD report. As experience grows it should be possible to develop an assessment scheme which, although using the existing "points to consider", aims to order the assessment process in a logical fashion. In the UK the ACGM is working towards a new version of its guidelines which will include the identification of the individual steps to overall decision making. Each step or question may be independent but should as far as possible be placed in a logical sequence with other steps. A decision needs to be reached under each step and each step may consist of a number of points to consider. An example might be:

Step/Question

Is it reasonable to expect the organism to move away from the release site?

Points to consider:

- survival forms
- pollination characteristics (if a plant)
- nature of release site (geography/metereiological conditions)
- expected effects of genetic manipulation on survival forms
- genetic transfer possibilities

Step/Question

It is reasonable to believe that the organism could multiply in the environment -- either at the release site or elsewhere?

Points to consider:

- growth and survival characteristics
- susceptibility to temperature, UV, desiccation
- expected effects of genetic manipulation on survival forms
- DNA transfer possibilities

In building up a risk assessment scheme based on a logic path the ACGM intends to improve assessment of release projects in a number of ways:

1. to develop consistency of approach
2. to assure completeness of review -- the scheme should ensure that all points of concern are considered
3. to identify areas where, in a particular case, more data are required
4. to amplify the current "points to consider" documents
5. to encourage scientific consensus
6. to give a more easily understood basis to the review process -- a point of particular value in the vital area of public confidence

Overview

Perhaps the greatest threat posed by the application of this new technology in the environment is uncertainty -- we cannot after all give absolute guarantees in this or any other area of endeavour. This uncertainty is enhanced in the case of micro-organisms by their ability to multiply and to provoke wide concerns.

The best way forward is to get the regulatory balance right and to ensure that there is widespread confidence that any potential for harm is being properly considered and taken account of and also that, if necessary, control measures can be enforced.

Given that genetic manipulation involves both promise and threat (or promise and uncertainty), what do we do about it? The answer must be to make sure that we have a national system that:

1. has the capability to keep out the unscrupulous and to stop activities that really do pose a threat, i.e., effective controlling legislation;
2. gets the balance of advice right, preferably harmonised on an international basis, so that the opportunities are not lost but the threats are dealt with;
3. involves an appropriately constituted "watchdog" body that can weigh up the concerns and give advice to those who have the controlling legislation and to those proposing to do the work;
4. instills public confidence -- perhaps the best way of doing this is to get the first three points right and to deal with the problem in an open way, taking steps to educate the public about our measures for assessment and control

References

1. Recombinant DNA Safety Considerations OECD Paris 1986 ISBN 92-64-12857-3.

2. Coordinated Framework for Regulation of Biotechnology US Federal Register, June 26, 1986.

3. Guidelines for protection against risks posed by new in-vitro combinations of nucleic acids (5th revision), Federal Ministry for Research and Technology, Bonn, October 1986.

4. Draft Guidelines for DNA processing in the environment Ministry of Housing, Physical Planning and Environment, Netherlands, January 1987.

5. Guidelines for Industrial Application of Recombinant DNA Technology, Ministry of International Trade and Industry, Tokyo, Japan, June 1986.

6. The Planned Release of Genetically Manipulated Organisms for Agriculture and Environmental Purposes. ACGM/HSE/Note 3, Advisory Committee on Genetic Manipulation, Health and Safety Executive, London UK, April 1986.

7. Nature 1986, 496 323.

8. GMAG Note 14 - Revised Guidelines for the Categorisation of Recombinant DNA Experiments, London 1982.

REGULATION AND RISK ASSESSMENT FOR ENVIRONMENTAL RELEASE OF GENETICALLY ENGINEERED MICROORGANISMS IN DENMARK

M. Laake
Water Quality Institue, Academy of Technical Sciences
Biotechnology & Microbiology Section
11, Agern Allé, DK-2970 Hørsholm, Denmark

Introduction

The Environment and Gene Technology Act (Act no. 288) was passed by the Danish Folketing on June 4th, 1986, following 3-4 years of intensive public debate of the issues (1). The law is administered by the Minister of the Environment, and is enforced through the National Food Agency.

Since then, this administration has rapidly built up its expertise and capacity on human health and environment problems related to accidental and planned releases, engaging some 30 persons of scientific and other academic backgrounds.

The Water Quality Institute has been contracted by the National Food Agency to evaluate issues of environmental safety, ecological impact, monitoring and risk assessment of genetically-engineered microorganisms. A comprehensive literature survey was completed last April (2). We also analysed the needs for future research through a second report (3), proposing a joint research and development programme to be launched, which now has materialized into a Center of Biotechnology and Safety.

The Water Quality Institute will presumably continue to play the central role as advisor to both authorities and industries on environmental questions. A research group on microbial ecology and environmental biotechnology, which will be funded basically by two governmental research programmes, is presently beeing set up.

NATO ASI Series, Vol. G18
Safety Assurance for Environmental Introductions of Genetically-Engineered Organisms
Edited by J. Fiksel and V. T. Covello

The Law Situation

As the first country in the world, Denmark enforced in 1986 a law regulating gene technology (1). The purpose of this act is to protect the environment, nature and health, including considerations of nutrition in connection with the application of gene technology. Gene technology is defined as genetic engineering, including selfcloning and deletion as well as cell hybridization.

Genetic engineering is further defined as the technique by which heritable material, which does not usually occur or will not occur naturally in the organism or cell concerned, generated outside the organism or the cell is inserted into the said organism or cell.

The act states that great weight shall be attached to the character and ecological conditions of the environment as well as the risk for an undesired effect caused by the application of gene technology. Moreover, importance shall be attached to the protection of the population against health risks.

It is up to the Minister of the Environment, however, in detail to lay down rules in pursuance of this act, and in doing so he has to consult with several other ministers. A set of guidelines has been issued, listing points to consider for application, approval, safety and control of industrial, large-scale operations. These are to a large extent based on the recent OECD guidelines on biotechnology and safety, the GILSP rules.

The act itself lays down a fairly wide and flexible, but comprehensive framework for future regulation, concerning approvals, supervision, authorities, complaints and provisions for penalty. Both research, industrial production and deliberate release is included. It states that research has to be carried out in classified laboratories (§9). Production in which genetically engineered organisms or cells are generated or used shall not be commenced except with the consent of the county council with respect to discharge of genetically engineered organisms or cells into the environment (§10).

Deliberate release of genetically engineered organisms or cells, including deliberate release for the purpose of experiments, shall not be allowed (§11). The Minister of the Environment may in special cases give an approval of deliberate release, however, and in the first cases he has to consult with the Folketing itself. This is due to happen early in 1988 on the issue of recombinant oil crops, and on that occasion the politicians are expected through their discussions to guide the Minister in his future practise.

Any person who applies for approval for industrial production, deliberate release, substances, products or foodstuffs shall submit information and make examinations to elucidate the case, including examinations according to specific directions and at specific laboratories (§15).

The all together 37 paragraphs of the Environment and Gene Technology Act further details the conditions for approval etc., which will not be treated in this context. It is important to note, however, the comprehensiveness and flexibility built into the law, which allows adjustments in the enforcement to be made throughout the coming years. The Folketing has also stated that a revision is due to be made in 3 years' time, in recognition of the efforts being made by international bodies to arrive at a consensus on how to deal with the possible adverse effects to the environment.

Practical Experiences with Risk Assessment and Regulation

To my knowledge the number of applications for industrial production based on genetically engineered microorganisms amounts to approx. 5, of which 3 have been approved so far. One plant for the production of human insulin by *Saccharomyces cerviciae* got its permit and has been in operation since March this year by NOVO Industries. At least two applications for deliberate release of genetically engineered crop plants have been filed with the authorities, of which one is due for decision early next spring.

NOVO Industries are themselves conducting studies of growth rate and survival of the yeast in culture media, sea water, soil and sewage. So far the results indicate that the recombinant yeast grows slightly slower but survives as effective as the parent organism in all media. At realistic temperatures it survives well, with halflives of 6-8 days in sea water and soil. At high temperatures and in sewage water and sludge the halflives are in the range 0.2-3 days.

The recombinant cells are analysed by ELISA technique, with detection limits in the range of 1-10 CFU per ml or gram sample in most media. The organism has been detected quite frequently in the effluent, sludge and heat sterilized filter cakes from coarse filtration of the fermentation liquid, and in a few cases also in effluents from the product processing units.

In this case, the National Food Agency evidently was confident with the investigations performed by the company for approval. Although environmental data were quite scarce, these were supplemented by judgements made by national experts. In future cases a more systematic and comprehensive approach will probably be required.

Future Research on Environmental Satety and Risk Assessment

In 1987 the Danish government launched its 5-year biotechnology research and development programme, which presumably will allocate some 5-10% of the total budget of 500 mill. DKR for biotechnology and safety, environmental biotechnology, and other related subjects.

Environmental impact and risk assessment methodologies will be addressed by the Center for Biotechnology and Safety, organized as a co-operation between the National Food Agency, the Water Quality Institute, the Gene Technology Group at the Danish Technical University, and others. Several connections to university research groups will also be established through Ph.D. students.

The budget for this 3-year programme has so far not been settled with the Government, but the proposal amounts to some 45 mill. DKR, including 50% allocated from other sources. 11 research projects have been proposed, covering the following main areas:

- Detection methods for genetically engineered microorganisms
- Methods of biological containment
- Models for spreading and establishment
- Test methods and studies of survival, establishment, genetic transfer and effects
- Risk assessment methodologies and case-studies

The main emphasis of the programme is to educate young Danish scientists in the field, as well as to establish the necessary methods etc. in the relevant laboratories. In this context exchange of scientists with other countries is most welcomed and highly necessary. An extensive co-operation with U.S. EPA laboratories and some university groups has already been established. Further information on the research programme can be achieved from the author.

References (No. 2 and 3 are written in Danish)

(1) Ministry of the Environment (1986). The Environment and Gene Technology Act. Act no. 288 of June 4th, 1986.

(2) Laake, M. 1987. Genetically engineered microorganisms in the environment. Fate and effects. National Food Agency, Publ. no. 143, May 1987.

(3) Laake, M. 1987. Analysis of research and development needs for risk analysis of genetically engineered microorganisms in the environment. Water Quality Institute, Report no. 9623/237, May 25th, 1987.

EVOLUTION OF RECOMBINANT DNA GUIDELINES IN JAPAN.

Hisao Uchida
Teikyo University
2-11-1, Kaga, Itabashi-ku
Tokyo 171, Japan

I would like to describe here a brief history of the regulation of recombinant DNA technology in Japan. Since the recombinant DNA guidelines are based on our experiences on rDNA researches on the one hand, and philosophical as well as social considerations on the other, I shall start from overviewing evolution of rDNA research guidelines in Japan as well as mechanisms and factors in Japan which had influenced or had to be taken into account in formulating rDNA guidelines. I am in a sense qualified to attempt the overview, because I have participated in the initial drafting of the guidelines for rDNA experiments issued in 1979 by the Ministry of Education, Science and Culture (MESC), and by the Science and Technology Agency (STA), and have proposed subsequent amendments of the guidelines as the chairman of the Committee on Genetic Manipulation (CGM) which is supported by a League of 17 Biomedical Societies in Japan. I have also involved in the formulation of industrial rDNA guidelines as the chairman of the committee on recombinant DNA technology advising MITI (the Ministry of International Trade and Industry). Since 1983 until the end of the last March, I have been the President of the Molecular Biology Society of Japan. Nevertheless, since I am not a regulative official, my talk will represent purely personal views and comments on the evolution of rDNA guidelines in Japan. Moreover, as Daniel Koshland has once remarked, "although scientists do not wish to do harm, and regulators do not wish to stifle progress, and yet their differing needs and desires inevitably make the scientist hostile to control and the regulator conservative about progress". Therefore, my talk may unintentionally sound as biassed because of my stance as a scientist.

NATO ASI Series, Vol. G18
Safety Assurance for Environmental Introductions
of Genetically-Engineered Organisms
Edited by J. Fiksel and V. T. Covello

According to Jim Watson, the Asilomar Conference have cried wolf without having seen or heard one. The wolf that everybody had imagined at that time was not an ordinary wolf, which we could fight away with a club, but a creature so formidable that it could eat us up before we could even cry for help. In retrospect, it might have resulted from a loose imagination, as Dr. Watson has commented, but there was no other choice at that time except to draft experimental guidelines which resulted in the first NIH rDNA Research Guidelines published in 1976, as everybody here knows. Once regulatory measure has been taken by a US governmental agency, even this was in the form of "guidelines", immense impacts upon international burocracy were inevitable, and Japan was not an exception. Morever, it usually takes about three years time since something happens in the US, books are published describing "all about" the incident, they are then translated into Japanese, and until substantial number of copies are sold and read in Japan to build up public concern.

MESC and STA Recombinant DNA Research Guidelines

Introduction of a system from the US is always a prudent choice to secure social acceptance and support to start something new in Japan. In fact, the first MESC rDNA research guidelines were formulated by following, as far as possible, the 1976 NIH guidelines. Accordingly, the MESC guidelines suddenly introduced into Japan various new ideas and mechanisms, including the concept of guidelines themselves, biosafety committee and officiers, various containment levels and so on. Certain confusions were inevitable. Most importantly, it should be noted that there was at that time, and still is presently, no biosafety regulations for handling human pathogens in Japan. Patients are controlled but not pathogens, and the situation is not expected to change in the near future. Nevertheless, the present day Japan is reasonablly clean and hygienic, enjoying a very high level of longevity.

Therefore, introduction of the rDNA guidelines was possible only at the expense of singling out the recombinmant DNA technology from other biomedical convensions and practices. In another words, the rDNA guidelines in Japan were required to be self-sustaining or self-consistent without support from regulations on biohazard in general. Moreover, since the concepts of physical as well as biological containments were new to the biomedical field, young governmental officials who were assigned to draft the first regulations were experts in oversighting design and managements of nuclear plants. The situation inevitablly supported more stringent attitude towards the regulation of the rDNA technology from administrative point of view.

Although molecular genetic studies of procaryotes were well advanced in Japan at that time, rDNA cloning of eucaryotic genes was yet to become popular among most of universities and industries, and there was no major rDNA controversies nor complaints towards DNA regulations in Japan. Thus, the MESC guidelines provided justifications for setting up institutional biosafety committees consisting of members from broad sectors, sometimes from outside the institution. Universities in Japan at that time were not completely free from afterdamps of the previous student uprising.

It took about two years before the MESC guidelines were first drafted until they were issued in 1979. The MESC draft had to be reviewed and authorized by various Committees and finally endorsed by the Science Council of the MESC, which meets twice a year. Based on the advise of the Council, the Ministry finally publishes the guidelines, or their amendments. Guidelines could only be introduced into the Japanese administrative system by going through all these complicated burocratic manuvours, and thus restricting the flexibility of the guidelines. In fact, just before the first MESC guidelines were published, the NIH guidelines were already relaxed extensuvely, reflecting the declining anxiety about the hazards of rDNA manipulations. However, the change could not be taken into account in the first MESC guidelines in time. Therefore, the MESC guidelines were already obsolete when they were first published.

The Council for Science and Technology was established in the Prime Minister's Office as the supreme advisory body to the Prime Minister in formulating and promoting the policies of the Goverment for science and technology. The administrative tasks of the Council are to be handled jointly by STA and MESC. However, activities of the Council is presently limited by the capacity of STA. The Council has a Panel on Life Science which in turn has a Sub-Panel on rDNA. The Sub-Panel is responsible to draft guidelines and also to review research proposals involving rDNA technology which are not under supervision of MESC. Research activities in private sectors are, therefore, covered by the STA guidelines. The first rDNA Research Guidelines were published in 1979, conforming with those published by MESC.

Above chronology clearly indicates that the legislative philosophies in formulating "rDNA guidelines" in Japan and the US are different. In Japan, the guidelines are authoritative directions deliberatively published by the Government and difficult to change from within, although they stated clearly that guidelines are to change when our accumulating knowledge demands. On the other hand, NIH previously revised their guidelines almost every 3 months, and they state that their guidelines cannot anticipate every possible situation, and will never be complete or final. No mechanism is apparent in MESC to accomplish this much flexibility. In fact, the MESC guidelines do not include procedures for future evolution of the guidelines. Until very recently, the MESC Committees are reluctant to take initiative themselves to draft or propose amendments of the guidelines without appropriate solicit from scientists outside the Ministry.

17 major Biomedical Societies in Japan, including the Molecular Biology Society of Japan, the Japanese Biochemical Society, the Japanese Society of Bacteriology, the Japanese Society of Virology, the Japanese Cancer Society, the Japanese Society of Agricultural Chemistry, the Japanese Society of Fermentation Industry, formed a league to support a voluntary body called the Committee on Genetic Manipulation (CGM) which in turn has proposed to the Government necessary improvements of the guidelines and expressed other opinions, from scientific viewpoints. Since its establishment in 1978, CMG has played a very influencial role in the evolution of rDNA guidelines in Japan. However, the major revision of the MESC guidelines in August, 1982 took a year and a half after initiative proposal of the revision made by CGM in February, 1981. I myself was

excluded from revising tasks of the MESC guidelines, because I was the one who proposed the revision.

The rigid guidelines have been expensive to MESC to support rDNA researches: by 1985, 48 P3 laboratories were constructed in 24 universities and 9 research institutions, although less than 3% of all rDNA projects were carried out in P3 facilities during 1985. No P4 laboratory was constructed under MESC control; researches involving rDNA technology are in progress in 111 universities and institutions in Japan, according to recent statistics disclosed by MESC. Therefore, the rigid guidelines forced the Government to increase tremendously the budgets for DNA researches in biomedical fields. Annual trends in the number of rDNA research projects approved by the MESC Biosafety Committee were the following:

Fiscal Year	Number of Projests
1979	95
1980	221
1981	375
1982	439
1983	76
1984	65
1985	115

A sudden drop in 1983 reflects the major revision of the MESC guidelines in the previous year, which appointed the head of each institution to review and approve institutional proposals of covered experiments before their initiation. Now, the MESC biosafety committee reviews only four categories of proposals which require specific review by the committee: (1) Use of uncertified Host-Vector Systems. (2) Experiments involving cloning of genes coding for the biosynthesis of protein toxins for vertebrates. (3) Experiments involving infection or inoculation of whole plants or animal by recombinant DNA organisms. (4) Experiments involving intentional (or deliberate) release of recombinant DNA organisms into environment. Therefore, except for experiments exempt from the guidelines, every rDNA project presently carried out in universities and institutions supervised by MESC are reviewed by either institutional or MESC biosafety committee before initiation. The total number of rDNA projects are still increasing rapidly.

It is pertinent to go into some details of the above categorization and to discuss characteristics of the current rDNA guidelines regulating experiments carried out in Japanese universities and institutions under MESC. (1) HV Systems certified by MESC are: EK1, EK2, SC (Saccharomyces cerevisie), BS (Derivatives of B. Subtilis 168 sustaining multiple nutritional requirements) and cultured cells (both animal and plant cells). EK2 is the only HV2 system. Therefore, a considerable number of other host-vector systems were proposed and approved for use by specific applicants. Under the current legislatory practice, certification of a new HV system for general use require approval by the Minister based on the advice of the Science Council of the MESC. (2) Toxins. NIH guidelines say "potent toxins", but MESC says "protein toxins". Several proposals in this category were made by medical bacteriologists. (3) The MESC guidelines regulate production and handling of "live cells" containing recombinant DNA molecules made by in vitro reactions. Recombinant DNA molecules themselves as well as recombination in vivo are exempt from the regulation (Natural exchangers are exempt from guidelines.) In spite of the understanding, MESC is presently reviewing all experiments involving whole insects, animals or plants. The main point to be checked is whether the experiment in question will be carried out under "reasonable" containment or not, irrespective of possible or impossible adverse effects. This unconditional containment policy circumvented and obscured the more important question of of how novel a recombinant organism is. In MESC guidelines or any other Japanese guidelines, no statement corresponding to III-B-4 of the NIH guidelines which describe information on "Experiments Involving Whole Animals or Plants" is included. The RAC/NIH is to review proposals only if fragments greater than two-thirds of a eukaryotic viral genome are to be used, or if recomninant molecules are to be deliberately transfered into human subjects. The Frankenstein syndrome is still obvious in Japan, and I hope AIDS will not stay and proliferate in Japan that long. As far as transgenic or infection experiments involving whole animals or plants are regulated as presently reviewed, no discussion has started openly on "deliberate" release of recombinant organisms into environment. In fact, no proposal was made which fall into the category (4) above.

Publication of OECD/CSTP Considerations

The international initiative taken by OECD/CSTP to harmonize the guidelines for industrial, environmental and agricultural applications of rDNA technology very much encouraged implementation of recombinant DNA guidelines at various Ministries other than MESC and STA. These Ministries have been keen about future development of recombinant DNA techniques which will open up promising possibilities in a wide range of applications.

After four years of study and preparations, MITI has issued in June, 1986 new Guidelines for industrial application of rDNA technology. The guidelines are consistent with the OECD report on Industrial Application of Recombinant DNA Technology, which endorsed the concept of Good Industrial Large-Scale Practice (GILSP). Accordingly, STA hastily amended their "research" guidelines in August, 1986 by relaxing allowable fermentation volumes and extending the range of restricted hosts. They published a list of 13 bacteria which are "certified" if used to host DNA obtained from procaryotes, lower eucaryotes and their viruses which are not included in the lists specifying organisms of noticable risks. Nevertheless, since too much patching up of the existing guidelines has obscured the original intent of the Council for Science and Technology, complete revision of the research guidelines are under discussion. At the same time, the Agency started discussion on deliberate releases.

The Ministry of Health & Welfare (MHW) has issued the guidelines for production of pharmaceuticals and biologics using rDNA technology, which are generally consistent with the OECD/CSTP report as well as with the MITI guidelines. Ministry of Agriculture, Forestry & Fisheries (MAFF) started feasibility studies of setting up rDNA guidelines. They are naturally interested in agricultural and environmental applications of biotechnology.

In December 1986, MAFF published their draft guidelines for public comment. In the draft, they introduced the concept of Good Outdoor Practice (GODP) to describe a controlled outdoor environment for limited field testing.

According to the draft, GODP is "application of rDNA plants in the restricted area so that the rDNA plant can not reproduce naturally outside the area and its pollen and spore cannot influence genetically to the plants grown outside the area." Detailed specification of GODP is yet to be discussed. The draft also mentioned applications of rDNA microorganisms in the controlled model environment; "Applications to the restricted area of rDNA microorganisms intended to be used in the open environment, where the dispersion of rDNA microorganisms and genetic material might be minimized." The Ministry is not decided yet when to implement the guidelines. Environment Agency of the Japanese Government has started, since 1982, considerations and preliminary research on how to evaluate environmental impacts of biotechnology including deliberate environmental releases of engineered microorganisms, or unintentional releases into environment by accidents. Although it is still too early to foresee when they could conclude any policy, one interesting observation they made was that certified hosts (EK1 or S. cerevisie) suspended in environmental soil extracts or water from rivers die much faster when extracts and water were not sterilized before inoculation of test organisms. However, they have not found out what is affecting the survival, nor how general is the observations. This is one of difficulties to deal with "environment".

MESC is also presently considering another major amendment of the present rDNA research guidelines, which have to be updated. Although the MESC guidelines cover only research activities, their coherency with guidelines covering practical applications is considered to be necessary. One possibility will be to regulate preparation of recombinant organisms separately from the use and handling of specific clones, because the former may involve handling of pathogenic donor organisms.

Thus, the OECD/CSTP activity has been very influential to promote drafting of rDNA guidelines for practical as well as research applications. The problem is that there is in effect no Governmental agency in Japan which is mandating coordination among Ministries as far as the rDNA guidelines are concerned. Obviously, there are two prototypes of rDNA guidelines in Japan, one evolved out of the NIH guidelines, and the other implemented OECD/CSTP considerations. Probably, coordination and harmonization of the context of the guidelines are not the major problem, as I suspect. This may be because coordination and harmonization are the virtue of the Japanese. Instead, each Ministry appears to be much more concerned about claiming by the name of rDNA technology the maximum extent of a new territory under his control. Therefore, there is some confusion among industries in Japan to which Ministry they should apply their proposals for a review.

As the chairman of the Committee on rDNA Technology which has drafted for MITI the first industrial guidelines in Japan, I would like to mention a few words about the MITI guidelines: First of all, I acknowledge OECD ad hoc Group, chaired by Dr. Roger Nourish, assisted by OECD Secretariat Miss Bruna Teso, for providing in time the Report on Recombinant DNA Safety Considerations, to have given an international background to publish industrial guidelines now in Japan, and to have indicated frameworks to be implemented in the MITI guidelines. And because we could have the opportunity to participate in the activity of the ad hoc Group to reflect our own ideas, experiences and history into the final report.

Guidelines in Japan are not easy to revise or amend as we have seen for the rDNA research guidelines. Therefore, the MITI guidelines describe a framework and ideas, as well as most of the considerations as listed up in the OECD/CSTP report. Instead of classifying every possible situations according to risks, the guidelines state only that the safety evaluation of each proposal will be made on case-by-case basis. Use of the word "containment" was avoided, since the word has been used frequently as almost synonimus to the safety. "Levels of containment" is the phrase difficult to translate into Japanese without giving impression of a self-contradiction, or a misconception that containment is more important than safety. Also the use of the word "environment" was minimized in the MITI guidelines.

We welcome the OECD report to have recommended, for the large-scale industrial application of rDNA technology, utilization of micro-organisms that are intrinsically of low risk, which can be handled under conditions of GILSP. The MITI definition of the GILSP host miroorganisms insists on non-pathogenicity, and either to have extended history of safe industrial use or to be able to grow under special cultural conditions, but otherwise the growth has to be restricted. More specifically, we are satisfied to find the OECD report to have mentioned self-clones and natural exchangers as examples of the GILSP category **unless they are pathogenic.** Self-clones and natural exchangers are therefore to be handled not as exemptions from the guidelines, but as candidates for GILSP organisms. Remember we do not have regulations on human pathogens. It is important for MITI to know what is going on in the Japanese industries, and to see if any industrial activity involving rDNA technology poses unusual occupational risks or not.

Under the MITI guidelines, recipient organisms are classified into the following five categories:

1) GILSP
2) Category 1
   Non-pathogenic, but not included in GILSP.
3) Category 2
   Pathogenicity to human is not deniable, but infection will not result in a serious illness. Effective preventive and therapeutic measures are available.
4) Category 3
   Human pathogens not included in Category 2, but effective preventive and therapeutic measures are avilable.
5) Special class
   Human pathogens without known preventive or therapeutic measures.

Another important point the OECD/CSTP Consideration clarified is the risk assessment of rDNA organisms: the Considerations state that "Since all but a small fraction of the genetic information in the modified organism is that of the recipient organism, a description of the recipient's properties provides initial information useful in assessing the properties of the organism derived by rDNA techniques. Information describing the difference between the properties of the modified organism and of the recipient organism defines the framework for safety assessment." As expressed as an international consensus, the statement was valuable in risk assessment of the modified microorganisms.

Since MITI issued industrial guidelines in June the last year, domestic industries requested the Minisitry to authorize their commercial production of more than 60 items designed by procedures involving rDNA technology. They were accordingly evaluated to be of low risks, and the Minister approved their handling under GILSP conditions. These were only a start, and I hope the MITI will be kept busy in responding to more and more proposals in years to come. However, the time will come to realize that, as noted in the OECD/CSTP Considerations, the requirement for a case-by-case judgement is not intended to imply that every case will require review by a national or other authority since various classes of proposals may be excluded.

Agricultural and Environmental Applications

As compared to the develpment of fermentation industry in Japan, activities involving agricultural or environmental applications of recombinant DNA technology are still premature at the present stage. However, "agriculture" cannot exist without genetic breeding to improve useful plants and cattles, and environmentally releasing it to cultivate and breed. Introduction of a new species into Japan is controlled strictly by quarrantine as is done in other countries. Unlike human pathogens, plant and animal pathogens are regulated by agencies of the Japanese Government. Therefore, I personally believe that there is no theoretical need to introduce a new regulatory measures controlling environmental releases of organisms because they were modified by rDNA technology. Rather, rDNA organisms should pose only a very small "incremental" difficulty over non-rDNA organisms to assess the risks, because, as stated in the OECD/CSTP Considerqations, "genetic changes from rDNA techniques will often have inherently greater predictablity compared to traditional techniques, because of the greater precision that the rDNA technique affords to particular modifications". Therefore, the main argument will be to realize how novel is a particular recombinant organism from DNA point of view.

On the other hand, because of overpopulation, the Japanese public in general is rather reactionary to the concept of "environmental releases".

Moreover, debates raised in the US on environmental releases of frost microbes and others are also influencing public opinion in Japan. At the same time, since the OECD/CSTP considerations suggested a provisional approach incorporating independent case-by-case review of a proposal before application, it is possible that the people concerned will prefer again singling out recombinant DNA technology from conventionl ones. In that case, we have to adopt much softer terminology like "limited application" or "GODP", and have to proceed a step by a step until the benefit without involvement of any risk becomes widely accepted.

It may be pertinent here to recall past experiences of environmental release of microbial pesticides in Japan. Although a number of microbial pesticides has been registered in Japan, and outdoor experiments and commercial sales of preparations have been permitted, the use of microbial pesticide in Japan is very limited and evaluation of its cost-benefit is still controvertial. The first case of microbial pesticide is Bacillus thuringiensis preparations. The bacteria was first isolated and described in Japan already during 1901 as a potent pathogen of the silk worm, and this fact has been and still is influencing greatly the evaluation of the bacteria as a microbial pesticide, although variants of the bacteria with low toxicity for the silk worm are isolated. I would like to remind you that, until say 50 years ago, silk was the main item of export from Japan. When the possibility of using B. thuringiensis as microbial pesticide was explored and became apparent in the US, the Japanese Government prohibited in 1959 its import to protect sericulture. Although B. thuringiensis preparations were registered as a Biological-Biorational Pesticide by EPA in 1960, the import was banned until 1971. Representative B. thuringiensis preparations have been produced in the US, France and the Soviet Union. In the US, B. thuringiensis preparations are exempted from tolerance requirements, and are applied at recommended doses even up to the day of plant harvest. In Japan, many companies applied for registration of B. thuringiensis preparations during 1974, but registers were granted in 1982 after waiting for another 8 years. Therefore, much data have been accumulated on the safety and persistence of B. thuringiensis preparations, paying particular attention to their effects on the sericulture. In the meantime, technics of the sericulture including disinfection procedures, advanced very much, and *B. thuringiensis* is no threat any more to cultivate silkworms. However, *B. thuringiensis* is listed in the current rDNA guidelines as an organism requiring P2 physical containment when used as a DNA donor. It is understandable that most companies diminished their interest to develop or improve microbial pesticides, in spite of the fact that a number of chemical pesticides are still causing troubles, and regulations by law will not be relaxed. This case may be an example to show the difficulty of changing a public concept after it was once established.

Finally, let me say a few words on the "Risk Analysis Approaches for Environmental Releases of Genetically Engineered Organisms". I think that the safety can be endorsed by our history and experience, but not by a theory or by an analytical logic. No doubt, it is important to have a theory, but without knowing what to control and what controls what, formulation of any general theory is impossible. Therefore, I am not convinced with deductive approaches which attempt first to formulate generic guidelines in order to assess risks associated with a particular proposal. It is true that recombinant DNA techniques have been used for more than a decade, and experimental or industrial guidelines have been usuful, a wide spectrum of environmental variations almost preclude any generalization unless generic guidelines become extremely restrictive, or of very limited applicability. I believe it might be more prudent to consider carefully and thoroughly each specific proposal by adopting the case-by-case approach, and to accumulate enough experience on environmental introduction of modified organisms before we can state anything general. Then, most of arguments are political rather than scientific, and this was the reason why I have confined my talk only within the framework of regulations.

INTERNATIONAL REGULATORY ISSUES FOR MODERN BIOTECHNOLOGY

J. David Sakura
BBN Laboratories Incorporated
10 Moulton Street
Cambridge, MA 02238

INTRODUCTION

The development and application of the "old" and "new" biotechnologies holds considerable promise for a variety of commercial products in areas such as medicine, agriculture, chemical processing and manufacture. As a consequence an entire industry has emerged in recent years that is built around the commercialization of biotechnology derived products. However, as these products are being tested and introduced into the market, stringent regulatory requirements may be established by national regulatory bodies partly in response to public concerns over environmental risks associated with certain biotechnology products and processes. These national policies and regulations may have an adverse effect on the orderly growth of this fledging industry.

A number of recent reviews of national environmental policies, regulations and guidelines covering products produced by biotechnology indicate that some type of risk assessment will be required during regulatory approval of the commercial products produced by biotechnology. (1,2,3) As shown by the biotechnology regulatory matrix prepared by the U.S. Office of Science and Technology Policy, a number of existing U.S. environmental laws, regulations and guidelines are based upon statutes that require an assessment of risk at some stage of product development. (4) A similar conclusion is reached from reviews of biotechnology regulations at the international level.

NATO ASI Series, Vol. G18
Safety Assurance for Environmental Introductions
of Genetically-Engineered Organisms
Edited by J. Fiksel and V. T. Covello

(5,6,7) As a consequence, existing and newly developed policies and regulations, such as the recent OECD Ad Hoc Working Group recommendations (5), present the regulator with a significant task of translating general safety and risk considerations involving recombinant DNA organisms into a series of coherent and scientifically defensible policies. Depending upon how these regulations are structured and implemented, the future regulatory climate could have a significant impact on the commercialization of biotechnology products.

At this time a need exists on the part of national regulatory agencies for adequate tools to assess environmental risks. The following are some regulatory issues associated with environmental applications of biotechnology that need to be considered while developing appropriate risk assessment methodologies:

- Are existing laws and regulations adequate (and flexible) to address the potential risks to the environment and human health by products produced by biotechnology?

- Is the current level of understanding of the environmental risks associated with biotechnology products sufficient to develop a scientifically sound regulatory approach?

- In view of the various approaches in regulating environmental hazards, what are some of the methodological needs of national and international regulatory bodies, that if met, could enhance their capabilities to conduct adequate risk assessments?

- To what extent is national and international coordination of biotechnology risk assessment research and data collection feasible and how can this objective be achieved?

In this paper, a simple regulatory framework, consisting of several dimensions, is presented that could be used as a guide during discussions on biotechnology risk assessment methodologies.

PRODUCT LIFE CYCLE AND REGULATORY REVIEW

The first dimension of the framework consists of a biotechnology product life cycle which highlights junctures where regulatory review of processes and/or products are conducted. This dimension captures an important feature of biotechnology, namely, its potential applicability to an unusually broad set of products and the diversity of the potential markets. It is important to note that the specific approach taken by various national and international bodies to regulate biotechnology products differs considerably in scope, complexity, and degree of comprehensiveness. Nevertheless, a generally accepted objective of most of these approaches is the minimization of risks of undesirable consequences at the least cost or burden to society.

Table 1 shows the chain of activities leading to product commercialization, consisting of four general stages in development of a biotechnology product: research and development, pre-production, production, and product distribution. Each stage in the product life-cycle is accompanied by a set of regulations, guidelines, etc., which may require some type of risk assessment. The following is a discussion of the status of recent biotechnology regulations at selected stages of product development.

Research and Development Guidelines

Much progress has been made at regulating the research and development stage, largely by means of the use of safety guidelines for rDNA research. Over the years, the stringency

TABLE 1

BIOTECHNOLOGY PRODUCT LIFE CYCLE

Product Stage	Regulatory Issues
Research and Development	Guidelines for Experimentation Laboratory Health and Safety Containment Scale-up Notification and Reporting
Pre-Production	Safety and Efficacy Testing Product Registration Product Licensing Notification and Reporting
Production	GMP/GLP Inspections Environmental Emissions Worker Safety
Product Distribution/ Sales	Transportation Import/Export Environmental Impacts Notification and Reporting

of these research guidelines has decreased as the degree of perceived risk diminished. The relaxation of the guidelines has been accompanied by a significant increase in the number and phylogenetic diversity of approved host-vector systems. In addition, a gradual downgrading of assigned risk levels for specific hosts has occurred, resulting in an increasing number of exempt experiments. As a consequence, recent revisions in U.S. and French rDNA guidelines have placed increased responsibility for oversight and review into the hands of local biosafety committees, rather than remaining under the jurisdiction of a central regulatory body.

In Japan where rDNA research guidelines were considered very stringent, relaxation of the number of restricted hosts, combined with an increase in allowable fermentation volumes for rDNA experimental work, now permits production of products such as human interferon, insulin, and growth hormone to be produced by E. coli or yeast in large-scale fermentation facilities.

The basic scope of some rDNA guidelines has been expanded, most notably in the U.S. and the U.K., to consider the risks of certain experiments conducted outside the laboratory. These experiments involve planned release of genetically engineered microorganisms, plants and animals.

Over the past 10 years, rDNA research guidelines have generally been mandatory for those institutions receiving public monies for their research. Industry compliance, however, to the research guidelines has been largely voluntary. (8) On the other hand, in response to the public's concern over risks involving modified microorganisms used in environmental applications, revisions to the Danish and West Germany rDNA research guidelines may result in previously voluntary guidelines becoming legally binding. (9) The economic and competitive impacts of these proposed changes in the research guidelines have yet to be determined. These actions at the national level may delay or possibly undermine current efforts by the European Community to develop a uniform set of guidelines for planned release applications.

Scale-up and Downstream Processing

During this stage of product development, the regulatory review process is complicated by the large number of environmental, product and worker safety statutes, each covering some phase of the product life-cycle. Risk considerations play an important role in the current debate on the need for and the development of new biotechnology regulations at this stage. (10,11) For example, the U.S. EPA is mandated with the implementation of eight major environmental statutes, and may develop specific regulations that deal with these various aspects of environmental protection.

Planned Release of Microorganisms

During the past several years, the first series of planned release experiments in the U.S. and in the U.K. have been reviewed and approved. In some cases the actual experiment was delayed, possibly for years, as legal disputes were settled in the courts. In anticipation of these planned experiments, a number of guidelines for assessing the risk of proposed experiments and product applications involving planned release of genetically modified microorganisms were developed. In the U.S., the EPA and the U.S. Department of Agriculture (USDA) have published policies for environmental applications of biotechnology. (12) Within the context of the Toxic Substances Control Act (TSCA), policies and new rules covering "new" and/or pathogenic and non-pathogenic microorganisms are currently being developed by the EPA. Based upon these policies, a number of specific regulations covering plant pests (13) and "new" microorganisms (14) have been proposed or are currently under development.

Similar guidelines for assessing the risk of planned releases of genetically manipulated organisms have been issued by the U.K. Advisory Committee on Genetic Manipulation (ACGM). (15) In contrast to the voluntary ACGM guidelines, Danish law now requires government approval based upon the recommendation of a committee of experts of all deliberate release experiments. (10)

Although deliberate release of genetically engineered microorganisms has been expressly prohibited in Germany and Japan, both governments and their regulatory agencies have been developing planned release guidelines. For example, according to the Japanese Ministry of Agriculture, Forestry, and Fisheries (MAFF), guidelines for deliberate release of rDNA plants and possibly microorganisms will be developed.

In France, AFNOR (Association Francaise de Normalisation) in conjunction with the French Ministry of the Environment studied the issue of environmental release, and a report was expected by the end of 1986. It is possible that in France review of planned release experiments may be a joint effort between industry and government.

Although the recommendations of the OECD Ad Hoc Working Group may accelerate the establishment of a uniform regulatory review mechanism for environmental applications of genetically modified microorganisms and plants, it is likely that widespread approval of planned release experiments in Europe and Japan may be several years in the future because of a number of technical obstacles in assessing risk and possible adverse political factors.

AGENCY RISK ASSESSMENT/MANAGEMENT FRAMEWORK

The second dimension of the framework involves the process by which a regulatory agency addresses biotechnology risk assessment and risk management issues. Five discrete step in the agency's regulatory risk assessment/risk management process can be envisioned:

Risk Identification and Description

During this initial step, the agency identifies and selects the hazards (human health, safety, environmental) for which it has the statutory mandate to regulate. As noted by Brickman et al. (16) there are striking differences in the way American and European authorities go about this process. The U.S. approach is notable for its legal and procedural complexity, whereas the European approach is far less complicated. This complexity is due, in part, to the way enabling legislation is written in the U.S., which is often in schematic language, resulting in complex regulations and procedural requirements.

Risk Estimation or Assessment

This step involves an integration of the assessments of the risk source, pathways of exposure, extent of exposure, dose-response, etc. which leads to an overall estimate of risk. From a regulatory perspective, the U.S. approach towards regulating chemicals (and by analogy genetically modified microorganisms) often involves widespread use of generic decision-making rules, whereas the European approach involves an administrative review of substances on a case-by-case approach, and seldom employs formal risk assessment or cost-benefit analyses.

Risk Evaluation and Control Option Evaluation

As part of the risk management process, regulatory agencies will evaluate the risks and the control options available in order to achieve an acceptable level of risk. This process is often impeded, especially in the U.S., by the ambiguity in the acceptable level of risk as prescribed by the enabling statutes. For example, the EPA is to regulate those chemicals that pose a "significant risk" (TSCA S4(f)) or "unreasonable risk" (TSCA S5(f). Presumably these risk considerations will used when the EPA reviews environmental applications of genetically engineered microorganisms.

Risk Mitigation Decisions

Here regulators must often come to their decision by setting priorities and choosing the appropriate risk reduction action(s) that is consistent with their particular assessment of acceptable risk. As would be expected, these judgements may differ from country to country. In a study of international environmental laws by the U.S. Congressional Budget Office (17), these judgements appeared to be fairly consistent with regard to the levels of air and water pollution in the U.S., Canada, and West Germany. By contrast, Japan's air and water standards appear to be much stricter. To complicate the process, each country mentioned above uses various approaches (e.g., performance standards or other attainment strategies, monitoring and enforcement, outright bans, etc.) to reach these environmental objectives.

In contrast to the experiences of the chemical industry, those that will be potentially regulated, i.e., the biotechnology industry, have been cooperating actively with government officials in assessing risk and developing a risk mitigation policies. (18)

Risk Assessment Issues and Regulatory Needs

Table 2 summarizes selected agency regulatory needs at each step of the regulatory process. In general, these can be divided into two groups: short-term and long-term informational

TABLE 2

AGENCY SHORT/LONG TERM NEEDS BY REGULATORY STAGES

	Regulatory Stage		Agency Needs
1.	Risk Identification and Description	o	A list of key biological descriptors needed to adequately characterize genetically modified and donor organisms
		o	A list of key descriptors for interactions of modified organisms with ecosystems
		o	Standards for characterizing ecological, pathological, and physiological traits of organisms
2.	Risk Estimation (Assessment)	o	Adequate measures for describing the magnitude of harm to ecosystem and/or ecological dysfunction
		o	Temporal and spatial descriptors
		o	Identification or development of a marker gene
		o	Qualitative and quantitative descriptors for describing risks to humans and the environment
		o	Development of simulation models to describe survival and growth of microorganisms
		o	Methods to identify human and ecological populations at risk
		o	Development of dose-response models

TABLE 2 (cont'd)

Regulatory Stage	Agency Needs
3. Risk Evaluation	o Methods to assess risk cut-off levels
	o Determination of regulatory end-points
	o Data selection protocols
	o Establishment of formalized approach vs. case-by-case approach to risk evaluation
	o Identification of prerequisite scientific expertise
4. Control Options Evaluation	o Development of short-term and long-term test protocols
	o Development of large-scale and small-scale test methods
	o Development of monitoring and detection systems
	o Development of site decon tamination technologies
	o Measures to determine control option effectiveness
5. Risk Mitigation Options	o Development of appropriate databases
	o Mechanisms to assist in agency priority setting
	o Methods for balancing hazard(s) vs. economic burden

and methodological needs. An important need at the international level is the development of a database on microorganisms that would allow for comparing organism function and taxonomy as well as ecological responses. A second research need is to develop a predictive, quantitative risk assessment methodology similar to those currently used for chemicals. The proposed methodology would be useful in predicting the effects of regulatory mechanisms designed to control the survival and proliferation of genetically engineered microorganisms.

COORDINATION AND HARMONIZATION OF BIOTECHNOLOGY ENVIRONMENTAL REGULATIONS

A uniform set of regulations and coherent administration of these regulations by various regulatory bodies is critical for the long term success of the biotechnology industry. At the international level, a common European approach to biotechnology regulations and standards has been called for by both industry and governments. (19) To this end, several international groups have been organized to address the issue of coordination, including the OECD Ad Hoc Working Group, and the Biotechnology Regulations Interservice Committee (BRIC) composed of representatives from the various services of the European Commission. At the national level, the U.S. Biotechnology Science Coordinating Committee (BSCC) and the Interdepartmental Committee on Biotechnology (ICBT) in the U.K. have similar coordinating roles within their respective governments. One indication of the need to coordinate regulation of biotechnology has been the series of published guidelines (20) or regulatory matrices (12) designed to clarify the role of various agencies within the U.S and U.K. governments, and to identify the appropriate regulations and products that fall within their mandates.

Recent experiences by U.S. regulatory agencies (FDA, USDA, EPA) in pre-market review of biotechnology products for health and environmental applications suggest that coordination among governmental agencies can lead to expeditious review of product applications. For example, the first open-air field test of a genetically engineered plant containing a pesticide (*Bacillus thuringiensis* toxin) was approved by the USDA within two months after submission. (21)

However, despite these near term successes, national industrial policies such as the current U.S. initiative to strengthen its patent policy, national and international public interest groups, and jurisdictional issues among the national regulatory agencies may impede the harmonization process.

CONCLUSIONS AND FUTURE TRENDS

The following conclusions are given regarding the current international status of biotechnology regulations:

- The majority of biotechnology regulations and guidelines have been developed by various nations for regulating the earlier stages, e.g., research and development, and pre-market approval, of product development. Because of an unblemished laboratory safety record and a reduction in the preceived risk, rDNA guidelines for laboratory research will continue to be liberalized, and should promote more rapid commercialization of biotechnology.

- Besides new guidelines for large-scale fermentation, it appears unlikely that significant specific regulations covering biotechnology products at the production and distribution stage will be developed. Most countries will rely on existing regulations to cover these phases of the product life cycle.

- While progress has been made in the U.S. regarding agency jurisdiction over biotechnology products and processes, the results of similar efforts within the European Community and in Japan may be several years in the future.

- Public opinion towards biotechnology appears to be relatively high throughout the world. Nevertheless, ethical considerations associated with gene therapy, or some untoward environmental event involving a biotechnology product or process may lead to development of more stringent regulations of biotechnology.

REFERENCES AND NOTES

1. Arthur D. Little, Inc., "Review and Analysis of International Biotechnology Regulations - A Report Prepared for a Consortium of U.S. Government Agencies," May 1, 1986.

2. Mantegazzini, M., "Evaluation of the Risk for the Environment of Products and Processes in the Field of Biotechnology," Study Contract No. 84-B-6602-11-007-11-N, Final Report to the Commission of the European Communities, Directorate General of Environment, Consumer Protection, and Nuclear Safety, October, 1955.

3. Kureczka, J., "Biotechnology and Governmental Regulations: Overview and Recommendations," Public Affairs Report (University of California, Berkeley) 25, 1984, pp. 1-13.

4. U.S. Office of Science and Technology Policy, "Coordinated Framework for Regulation of Biotechnology," Federal Register 50, November 14, 1985, pp. 47174-47195.

5. OECD-AD-Hoc Group on Safety and Regulations in Biotechnology, Draft Report DSTI/SPR/85.23, 1985.

6. A preliminary inventory was prepared in draft form in March 1985, and reviews existing regulations covering human and animal health products, agriculture, environment and worker safety.

7. The draft document "An International Approach to Biotechnology Safety" was prepared for UNIDO by U.S. consultant, Geoffrey M. Karney.

8. Countries where voluntary compliance is observed by industry include among others: the U.S., U.K., Japan, Belgium, Federal Republic of Germany, France and the Netherlands.

9. Conzelmann, C. and D. Claveloux, "Europe Fails to Agree on Biotech Rules," New Scientist, July 10, 1986, pp. 19-20.

10. Arthur D. Little, Inc., "The Suitability and Applicability of Risk Assessment Methods in Environmental Applications of Biotechnology," V. Covello and J. Fiksel, eds., Report No. NSF/PRA 8502286, National Science Foundation, August, 1985.

11. American Society for Microbiology, "Engineered Organisms in the Environment: Scientific Issues," H.O. Halvorson, D. Pramer, and M. Rogul, eds., Washington, D.C., 1985.

12. U.S. Office of Science and Technology Policy, "Coordinated Framework for Regulation of Biotechnology," Federal Register 51, June 26, 1986, pp. 23302-23350.

13. U.S. Department of Agriculture, Animal and Plant Health Inspection Service, "Introduction of Organisms and Products Altered or Produced Through Genetic Engineering Which Are Plant Pests or Which There is Reason to Believe They Are Plant Pests," Federal Register 51, June 26, 1986, pp. 23352-23366.

14. The EPA has announced development of a "significant new use" rule for pathogens and microorganisms containing genetic material from pathogens and a rule requiring

submission of a Premanufacture Notification (PMN) prior to field testing of a "new" microorganism. (51 Federal Register, pp. 14543, April 21, 1986.)

15. U.K. Advisory Committee on Genetic Manipulation, "The Planned Release of Genetically Manipulated Organisms for Agricultural and Environmental Purposes - Guidelines for Risk Assessment and for the Notification of Proposals for Such Work," ACGM/HSE/Note 3, April, 1986.

16. Brickman, R., Jasanoff, S., and Ilgen, T., "Controlling Chemicals: The Politics of Regulation in Europe and the United States," Cornell University Press, Ithaca, N.Y., 1985.

17. Congressional Budget Office,Congress of the United States,"Environmental Regulation and Economic Efficiency," Washington, D.C., March 1985

18. "Safety and Regulation in Biotechnology: A Position Paper Prepared by the European Committee on Regulatory Aspects of Biotechnology (ECRAB)," Swiss Biotech 4 No. 5, p 15-20 (1986).

19. Creasey, P., "Prescriptive Policies for Industrial Biotechnology in Europe," in Industrial Biotechnology in Europe - Issues for Public Policy, D. Davies, ed., Center for European Policy Studies, 1986.

20. Interdepartmental Committee on Biotechnology, "Biotechnology, Support and Regulations in the U.K. - A Plain Man's Guide," The Laboratory of the Government Chemist, Department of Trade and Industry, April, 1986.

21. Crawford, M., "Test of Tobacco (Plant) Containing Bacterial Gene Approved," Science 233, September, 1986, pp. 1147.

NATO ASI Series G

Vol. 1: **Numerical Taxonomy.** Edited by J. Felsenstein. 644 pages. 1983. (out of print)

Vol. 2: **Immunotoxicology.** Edited by P.W. Mullen. 161 pages. 1984.

Vol. 3: **In Vitro Effects of Mineral Dusts.** Edited by E.G. Beck and J. Bignon. 548 pages. 1985.

Vol. 4: **Environmental Impact Assessment, Technology Assessment, and Risk Analysis.** Edited by V.T. Covello, J.L. Mumpower, P.J.M. Stallen, and V.R.R. Uppuluri. 1068 pages. 1985.

Vol. 5: **Genetic Differentiation and Dispersal in Plants.** Edited by P. Jacquard, G. Heim, and J. Antonovics. 452 pages. 1985.

Vol. 6: **Chemistry of Multiphase Atmospheric Systems.** Edited by W. Jaeschke. 773 pages. 1986.

Vol. 7: **The Role of Freshwater Outflow in Coastal Marine Ecosystems.** Edited by S. Skreslet. 453 pages. 1986.

Vol. 8: **Stratospheric Ozone Reduction, Solar Ultraviolet Radiation and Plant Life.** Edited by R.C. Worrest and M.M. Caldwell. 374 pages. 1986.

Vol. 9: **Strategies and Advanced Techniques for Marine Pollution Studies: Mediterranean Sea.** Edited by C.S. Giam and H.J.-M. Dou. 475 pages. 1986.

Vol. 10: **Urban Runoff Pollution.** Edited by H.C. Torno, J. Marsalek, and M. Desbordes. 893 pages. 1986.

Vol. 11: **Pest Control: Operations and Systems Analysis in Fruit Fly Management.** Edited by M. Mangel, J.R. Carey, and R.E. Plant. 465 pages. 1986.

Vol. 12: **Mediterranean Marine Avifauna: Population Studies and Conservation.** Edited by MEDMARAVIS and X. Monbailliu. 535 pages. 1986.

Vol. 13: **Taxonomy of Porifera from the N.E. Atlantic and Mediterranean Sea.** Edited by J. Vacelet and N. Boury-Esnault. 332 pages. 1987.

Vol. 14: **Developments in Numerical Ecology.** Edited by P. Legendre and L. Legendre. 585 pages. 1987.

Vol. 15: **Plant Response to Stress. Functional Analysis in Mediterranean Ecosystems.** Edited by J.D. Tenhunen, F.M. Catarino, O.L. Lange, and W.C. Oechel. 668 pages. 1987.

Vol. 16: **Effects of Atmospheric Pollutants on Forests, Wetlands and Agricultural Ecosystems.** Edited by T.C. Hutchinson and K.M. Meema. 652 pages. 1987.

Vol. 17: **Intelligence and Evolutionary Biology.** Edited by H.J. Jerison and I. Jerison. 481 pages. 1988.

Vol. 18: **Safety Assurance for Environmental Introductions of Genetically-Engineered Organisms.** Edited by J. Fiksel and V.T. Covello. 282 pages. 1988.

MIX
Papier aus verantwortungsvollen Quellen
Paper from responsible sources
FSC® C105338

If you have any concerns about our products,
you can contact us on
ProductSafety@springernature.com

In case Publisher is established outside the EU,
the EU authorized representative is:
Springer Nature Customer Service Center GmbH
Europaplatz 3, 69115 Heidelberg, Germany

Printed by Libri Plureos GmbH
in Hamburg, Germany